SEEDS
and their uses

SEEDS
and their uses

C. M. DUFFUS

School of Agriculture, University of Edinburgh
and East of Scotland College of Agriculture

and

J. C. SLAUGHTER

Dept. of Brewing and Biological Sciences,
Heriot-Watt University, Edinburgh

JOHN WILEY & SONS
Chichester · New York · Brisbane · Toronto

British Library Cataloguing in Publication Data:

Duffus, C. M.
  Seeds and their uses.
  1. Seeds
  I. Title II. Slaughter, J. C.
  633'.08'21   SB117   80–40283

ISBN 0 471 27799 1 (Cloth)
ISBN 0 471 27798 3 (Paper)

Typeset by Computacomp (UK) Ltd
Fort William, Scotland and printed at
The Pitman Press, Bath, Avon.

Preface

A few types of seeds form the main component of the diet of a large proportion of mankind. Even in the richer, or developed, countries where the diet is highly varied the population still depends heavily on the more or less direct consumption of a very limited range of seeds, e.g. as breakfast cereals, bread, and cakes. This range of seeds is extended only a little when seed products such as beer, chocolate, and cooking oils are included. It has also to be remembered that the large-scale meat production practised in many countries depends to a significant extent on the use of various seed meals and cakes as animal feedstuffs. Apart from their importance in nutrition, seed products are essential for many other purposes, e.g. drying oils in paints, glues, and pastes as well as likely raw materials in an oil-starved world. A considerable amount has been written about all these aspects as well as the purely agricultural uses of seeds, but most of the information is in expensive specialist monographs. The aim of this book is to provide a discussion of the properties and subsequent utilization of economically important seeds within a single volume. It outlines the growth habit and geographical distribution of the main cultivated seed-bearing plants. The biochemical and morphological changes accompanying seed formation are then described in detail, with emphasis on the deposition of the storage materials important in the nutrition of man and animals. After considering some of the problems relating to seed storage and the maintenance of viability and wholesomeness, the relationship between the chemical composition of seeds and the nutritional requirements of man and farm animals is explained. Finally, the processes used in the main seed-based industries are described with particular reference to the quality of seed required and the way in which seed properties determine aspects of the process and *vice versa*.

This book is based in part on lectures in applied biology given over the past few years at the University of Edinburgh. The second-year course is attended by students of agriculture, animal nutrition, botany, zoology, genetics, physiology, and crop and animal production. The book was written in response to student requests for a text covering the growth and uses of economically important seeds. With the success of the second-year course, a third-year course has just begun, with food production and food science as major components. The material in Chapter 4, in which the nutritive value of a range of seeds, notably cereal grains, legumes, and oilseeds, is assessed, forms the basis of the first part of this course.

We consider that this book should be of direct interest to students taking undergraduate courses in botany, biology, agriculture, nutrition, food science, and crop or animal science. Although the book requires an elementary knowledge of botany and biochemistry for a complete understanding, we hope that it may also be useful to anyone interested in learning about the properties and uses of seeds in the world today.

Many people have helped us at various stages, but we would like to express our particular thanks to A. H. Button, I. Fletcher, J. S. H. Hemingway, K. Inatomi, I. Longstaff, A. MacWilliam, P. R. Taylor, and G. A. R. Wood for providing useful information on industrial aspects of seed processing. We also thank M. P. Cochrane, J. P. F. D'Mello, J. H. Duffus, P. McDonald, and G. H. Palmer for comments and critical assessment of different parts of the manuscript.

C. M. DUFFUS
School of Agriculture,
University of Edinburgh

J. C. SLAUGHTER
Dept. of Brewing and Biological Sciences,
Heriot-Watt University

Contents

Chapter 1

The seed plants

1.1 GLOBAL ASPECTS

Seeds and their by-products are a major component of the diet of man. At world level, cereal grains contribute about 50% of the per capita energy intake (Table 1.1). Maximum reliance occurs in the Near and Far East, where over 65% of the total daily energy supply comes from cereals. In such developed market economies as North America and Western Europe the proportion is much less—around 25%. These figures are, of course, reflected in the overall sources of dietary energy, since in the former countries dietary energy comes mainly from vegetable products whereas in the latter much energy is derived from animal products.

Table 1.1. Percentage contributions of various food groups to daily per capita energy supply in developed and developing countries and the world in 1972–1974 (FAO 1977). (Reproduced by permission of the Food and Agriculture Organization of the United Nations)

Food group	1 Selected developing market economies	2 Asian centrally planned economies	3 Developed market economies	4 World
Vegetable Products				
Cereals	65.8	65.4	26.4	49.4
Pulses, nuts, seeds	7.2	7.0	2.7	4.9
Others (including roots, tubers, sugar)	20.8	18.4	34.5	28.3
Total vegetable products	93.8	90.8	66.6	82.6
Total animal products	6.2	9.3	33.4	17.4

1. Includes Bangladesh, Egypt, Ethiopia, Ghana, India, and Pakistan.
2. Includes China, Mongolia and Vietnam.
3. Includes North America and Western Europe.
The daily per capita energy supply (kilo joules) in the various categories is: 1, 8.49; 2, 9.58; 3, 13.97; 4, 10.67.

Other seeds, such as pulses, nuts, and oilseeds, are of particular importance in developing countries, mainly as a source of protein. In the Asian centrally planned economies of China, Mongolia, and Vietnam this group accounted for nearly 20% of the daily per capita protein supply in 1972–1974 with the remainder made up mainly by cereal grains (Table 1.2). This figure is much less for the developed market economies—around 5%—where the major part of dietary protein is in any case derived from animal products.

Table 1.2. Percentage contributions of various food groups to daily per capita protein supply in developed and developing countries and the world. (FAO 1977). (Reproduced by permission of the Food and Agriculture Organization of the United Nations)

Food group	1 Selected developing market economies	2 Asian centrally planned economies	3 Developed market economies	4 World
Vegetable Products				
Cereals	63.3	50.8	26.2	44.6
Pulses, nuts, seeds	14.8	19.2	5.1	10.9
Others (including roots, tubers, fruit)	7.3	9.3	10.1	9.4
Total vegetable products	85.4	79.3	41.4	64.9
Total animal products	14.6	20.7	58.6	35.1

1. Includes Bangladesh, Egypt, Ethiopia, Ghana, India, and Pakistan.
2. Includes China, Mongolia and Vietnam.
3. Includes North America and Western Europe.
The daily per capita protein supply (grams) in the various categories is: 1, 51.1; 2, 62.9; 3, 95.4; 4, 68.5.

The animal products which have replaced the foods of vegetable origin in the diet of those people living in the developed market economies are obtained to a large extent from farm animals whose diet contains varying amounts of seeds—notably cereals and their by-products, together with protein concentrates derived from oilseeds. Thus, the true—but indirect—consumption of seeds in these economies is very much higher than the figures shown in Tables 1.1 and 1.2 would suggest. The consumption of vegetable fats and oils, mainly derived from oilseeds, is not shown but generally accounts for between 4 and 10% of the total daily per capita energy supply.

Farm animals are large consumers of seeds and their by-products. Calves, pigs, and poultry, for example, depend on cereal grains as their main source of energy. The amount required depends on the animal's growth rate but can reach as much as 90% of the diet. Mature ruminant animals, such as cattle and sheep, also utilize seeds in the diet, but to a much lesser extent. Their function, in this case, is as a concentrated source of energy and/or protein.

The year-to-year changes in cereal production during 1961–1976 are shown for the developed and developing countries in Fig. 1.1, together with the corresponding trends in consumption as human food and animal feed. In the developed countries the margin between cereal production and its consumption as human food is large and increased from 331 million metric tons in 1961 to 544 million metric tons in 1974. The corresponding figures for the developing countries were 85 million metric tons in 1961 and 108 million metric tons in 1974. This margin is necessitated by the requirement for cereal grains as animal feedstuffs, and of course is much greater in the developed economies. In the developing countries the proportion of cereal grains used as animal feed is around 10–15% whereas in the developed countries it is around 55–60%.

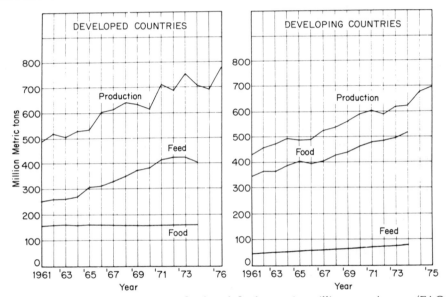

Figure 1.1   Cereals. Production, food and feed use—in million metric tons (FAO, 1977). (Reproduced by permission of the Food and Agriculture Organization of the United Nations)

Thus, seeds—and cereals in particular—are a dominating influence in world food production and account for over half of world dietary energy and protein supplies.

1.2 NATURE AND ORIGIN OF SEED CONSTITUENTS

The seed plants include nearly all of the crop plants and, with the exception of coniferous forests, constitute the dominating vegetation of most landscapes.

Most of these plants originate from seeds, which, after germination, develop into young plants or seedlings having roots, stems, and leaves. The plant grows, extending its root system into the soil and its stems and leaves upwards into the atmosphere. After some time, flowers form, fertilization follows and the seeds begin to develop and mature.

Mineral ions and water are taken in by the roots and conducted via the stem to growing tissues such as leaves. Carbon dioxide is incorporated into sugars during photosynthesis in the leaves and other organs. The substances generated, after biochemical transformations in the leaves and roots, are then transported to the developing seeds where lipid (fat), carbohydrate and protein are synthesized and stored.

The most striking biochemical features of the seed-forming plants is that they are largely composed of water and, with the exception of a few parasitic plants, have the ability to grow and reproduce, depending only on the gases in the atmosphere, the minerals in the soil and the radiation of the sun. The first feature is common to all living things but the second is restricted to the green plants with the exception of a few photosynthetic bacteria. Fig. 1.2 illustrates the major relationships between plants and animals on Earth. A chemical examination of plant material reveals that over 99% is made from the elements carbon, hydrogen, oxygen, nitrogen, phosphorus, and sulphur, with most of the remainder being accounted for by the chlorides of calcium, sodium, potassium, and magnesium. Many other elements, such as iron, zinc, copper, silicon, and iodine, occur in vanishingly small amounts. More sophisticated analysis of the dry material of plants indicates that usually over 85% of the major elements occur in the form of polymeric compounds: proteins, polysaccharides, nucleic acids, and complex lipids. In the case of the vegetative tissues of green plants, carbohydrate-derived compounds are by far the most dominant group. The single most common compound is usually glucose, either free, or more likely in combination as di- and trisaccharides, e.g. sucrose and raffinose, or as the polymers starch (α-1,4-linked) and cellulose (β-1,4-linked). Other hexoses, e.g. fructose and mannose, occur and may form important polymers in some plants. Always present to some

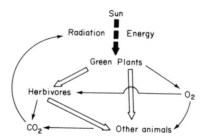

Figure 1.2   The main energy relations between green plants and animals on earth. Radiation from the sun supplies the energy for the formation of plant tissue from CO_2 and water with the equivalent release of oxygen. Animals derive their energy and tissue from the consumption of plants or animals with the concomitant uptake of oxygen and release of CO_2 and water. The radiation energy gathered from the sun by photosynthesis is eventually dissipated as heat arising from the metabolic activities of plants and animals

extent are a range of mixed pentose and hexose polymers—the so-called hemicelluloses. All of these compounds seem to have relatively defined functions in the plant: sucrose acts as a major transport compound, starch is usually a storage compound, whereas cellulose and hemicellulose occur in the cell walls and are the basic structural materials of plants. In the later stages of development of certain plants the structural role is supplemented by the phenolic polymer lignin.

The protein content of plants is usually low, although fairly diverse. Some protein is involved with lipids in forming membranes both around and within the cell but the largest fraction is probably the enzymic content of the cytoplasm. The exact composition of this depends, of course, on the degree of specialization of the cell, e.g. photosynthetic cells, transport cells and secretory cells, as well as on the general metabolic state of the plant, e.g. active growth, flowering or over-wintering.

As in most other organisms, DNA is the genetic material in plants and this material must be doubled before each cell division. More variable is the RNA content which, again as in other organisms, is involved in the process of protein biosynthesis. The amount of RNA therefore tends to depend on the metabolic state of the cell.

The term 'lipid' covers a wide range of compounds connected only by their solubility in organic solvents. The vegetative tissues of plants tend to contain only small amounts of lipids but a wide range of compounds such as glycerides, sterols, and phospholipids occur. In many cases these compounds have an essential function in cellular membranes. They are especially important in the chloroplasts where chlorophyll and the carotenoid pigments are responsible for the collection of the sun's radiation. Most plants also produce a range of low molecular weight lipids which feature in the essential oil, and in some cases highly distinctive lipids are produced, e.g. rubber (poly-*cis*-isoprene).

Seeds usually form a distinct biochemical contrast to the parent plant. Firstly, seeds are relatively dry and commonly contain only 10–20% moisture at maturity. The next variation is that, whilst most of the material of seeds is present as one of the common types of polymer, the seed polymers are almost certainly different from those of the parent plant. In seeds where the dominant compound is still carbohydrate this is likely to be starch with the β-glucans and hemicelluloses relegated to a minor level. In many seeds a major component is lipid, usually of the triglyceride type, and in seeds of this kind the protein content may be as high, if not higher, than the carbohydrate content. Seed proteins are characteristically different from the proteins of the parent plant. The amount of enzymic protein is low and the bulk of the protein is likely to consist of a few species of relatively insoluble proteins whose amino acid composition is rather restricted. These are commonly referred to as storage proteins.

All of these observations fit with the shape and function of seeds. Because of their compact shape most seeds have little need for large amounts of strengthening materials whereas, on the other hand, their function as non-photosynthetic plants in suspended animation requires them to carry a distinct reserve of easily digestible energy-rich material which can provide the raw materials and energy for germination.

An important point to remember about seeds is that, whilst a few polymers make up the bulk of the dry weight, the seeds may contain compounds in very low concentration which have an effect on animals. A major group here is the alkaloids, basic nitrogen-containing compounds of relatively low molecular weight derived, at least in part, from a few amino acids. Whilst most alkaloids have no physiological effects when eaten, many do and the range of response is wide, e.g. caffeine from coffee or cocoa produces only mild uplift whereas morphine from certain types of poppy is a potent pain killer which is extremely addictive. Compounds of this type can prevent the use of a seed for food or can, in fact, become the basis for its use. Seeds may contain many other, frequently poorly characterized, undesirable compounds such as toxic non-protein amino acids, anti-digestive factors, agglutinins, and lectins. The production of such compounds is usually species-specific or even variety-specific so there is scope for the plantbreeder to reduce or increase the content as desired.

The basic biochemical processes of biosynthesis in plants are similar to those in other organisms, but the range of reactions that occur is much wider in plants than in animals. This is because the plant has to synthesize all of its components from simple inorganic molecules whereas not only do animals contain a more restricted range of compounds, but they also cannot synthesize several of the compounds which they do use, and have to eat them ready formed in the diet. This is the biochemical basis for a whole range of essential compounds in animal and human diets, e.g. vitamins, essential amino acids, and essential fatty acids. A highly simplified diagram of the biosynthetic flow of carbon, hydrogen, and oxygen in plants is given in Fig. 1.3.

1.3 THE CEREALS

Cereal grains are the seeds of the cultivated grasses (*Gramineae*) which are grown for their edible starchy seeds. They include wheat, rice, maize, barley, oats, rye, grain sorghum, and various millets.

Wheat is now the major crop in temperate or dry climates and is grown extensively in the USA, USSR, Canada, and Australia. Its major role is in the production of flour for breadmaking. Rice is the preferred cereal in damp tropical climates, although if sufficient water is available it can flourish in such countries as Egypt, Italy and Australia. Maize (American corn) is more resistant to drought than either wheat or rice and is a popular cereal in many countries, including the USA, Brazil, Mexico, South Africa, and Argentina. In industrialized countries such as the USA with a well developed agriculture, a high percentage—up to 85%—of the crop may be fed direct to livestock,

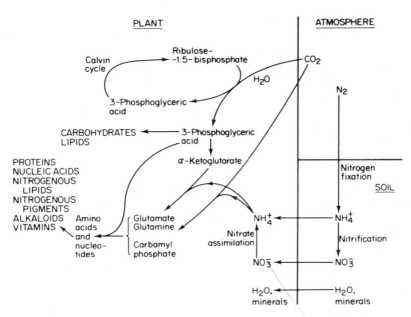

PLANT   ATMOSPHERE

Figure 1.3 Simplified outline of biosynthesis in green plants. The first organic carbon compound formed is 3-phosphoglyceric acid, whilst the earliest organic forms of nitrogen appear to be glutamate/glutamine and carbamyl phosphate. All the non-nitrogenous carbohydrates and lipids can be formed from 3-phosphoglyceric acid and all the nitrogenous components of the cell can be derived from glutamate/glutamine and carbamyl phosphate

otherwise it is used primarily as a food for humans. Barley is widely grown in almost all parts of the world and is mainly used in brewing and as an animal feedstuff. Oats, a hardy crop best suited to cool moist climates, is a major cereal crop, and a favourite feed for ruminant animals and horses. A small amount is used in the production of oatmeal. Rye, which grows well in poor soils and cold climates, is still grown extensively in Northern and Eastern Europe. It can be used for the production of bread and may be fed in restricted amounts to farm animals. Sorghum, a plant well adapted to dry conditions and high temperatures, is grown in the drier areas of Africa, India, China, and, increasingly, the USA, Australia, and South Africa. With the millets, which include several species of small-grained cereals also well adapted to dry conditions, it is eaten by many people in Africa and some areas of Asia and Latin America. The importance of these grains is becoming reduced as new drought-resistant varieties of wheat are introduced.

1.4 THE LEGUMES AND OILSEEDS

The physiology of growth and development in these plants has not been described as exhaustively as that of the cereals, possibly because they contribute considerably less to world food supplies. Certainly, the recent rapid rise in yields and production of wheat, rice, and maize, due to the introduction of new high-yielding crop varieties capable of responding to improved cultural practices, has meant that farmers are devoting a greater area of their arable land to cereals with a consequent decrease in the area devoted to food legumes. Little increase in grain legume production has been achieved in recent years, either by the development of new varieties or the introduction of new technology.

Cereal seeds have only one cotyledon or seed leaf and hence cereals are classified as monocotyledons. In contrast, legumes are dicotyledons, since their seeds have two cotyledons. In a cereal seed the testa is fused to the pericarp and hence the 'grain' is in fact a one-seeded fruit of the type known as a caryopsis, whereas legume seeds develop inside pods which may contain one or more seeds and vary greatly in size and structure.

One of the most studied legume groups is probably the food legumes or pulses. These include the common bean (*Phaseolus vulgaris*), garden pea (*Pisum sativum*), black gram (*Phaseolus mungo*), and red gram or pigeon pea (*Cajanus cajan*). Sometimes classified with these are the oilseed legumes, which include soybeans (*Glycine max*) and groundnuts or peanuts (*Arachis hypogaea*). Non-leguminous oilseeds include cottonseed (*Gossypium* spp.) coconuts (*Cocos nucifera* L.), oil palm kernels (*Elais guineensis* Jacq.), sunflower seed (*Helianthus annus* L.), linseed (*Linum usitatissimum* L.), and the rapes and mustards (*Brassica* spp.). The pulses, as a group, are of interest since they can generally be eaten after only minimal preparation such as soaking and boiling. They can be grown over a wide range of climates and soil types and thus make a direct contribution to the diet of many people in different parts of the world. Compared with the cereals, however, the average national yields of pulses have been very low and the gap continues to widen. One of the reasons for their low production is that many of the available varieties take a long time to mature, with the result that their growing season overlaps that of the cereals. Hence their daily productivity is low and the farmer must choose between the two groups of crops.

Soybeans are one of the world's most important sources of edible oil and protein and dominate world production of legume seeds (Table 1.3). Although they have been eaten in China for several thousand years they are not generally used for direct human consumption in the West. The greatest regional production outside the Far East is in North America, where they are an important raw material for the international food industry. For example, the extracted oils are used in the manufacture of margarine and the remaining oil cake can be fed to farm animals.

In terms of world tonnage of legumes produced, groundnuts are second only to soybeans (Table 1.3). Groundnuts can be grown in all tropical and subtropical countries. Major producers include India, Nigeria, and China. In general, they are not grown as a primary source of human food, but rather as a source of edible oil which is used in cooking or in the manufacture of

Table 1.3. Regional and World Production of Food Legumes for 1972 (Figures = 1 000 metric tons) (Milner, 1975). (Reproduced by permission of John Wiley and Sons)

Region	Dry beans (*Phaseolus vulgaris*)	Dry broad beans (*Vicia faba*)	Dry peas (*Pisum sativum*)	Chick peas (*Cicer arietinum*)	Pigeon peas (*Cajanus cajan*)	Cow peas (*Vigna* spp.)	Soybeans (*Glycine max*)	Groundnuts in shell (*Arachis hypogaea*)
Western Europe	609	809	262	130	—	6	—	21
Eastern Europe and USSR	321	25	5 405	2	—	—	810	3
North America	896	—	154	—	—	20	35 054	1 492
Latin America	3 851	167	107	186	34	—	4 038	1 451
Near East	186	347	10	235	—	10	20	508
Far East	3 908	3 410	4 331	6 530	1 548	27	12 740	8 495
Africa	1 237	428	392	332	66	1 081	25	4 534
Oceania	2	—	70	—	—	2	25	28
World	11 010	5 286	10 731	7 415	1 648	1 146	52 712	16 532

margarine. As with soybeans, the residue after oil extraction can be used in the feeding of farm animals (Section 4.3).

The greater part of world production of the non-leguminous oilseeds is crushed for edible oil. Inferior oil grades may be used for industrial purposes (Section 5.14) and the residues after oil extraction can be used as a high-grade protein supplement for farm livestock (Section 4.3.2). The cotton plant is grown in the USA, USSR, China and India, and is probably the most important of non-leguminous oilseeds. The oil palm requires a tropical rainy climate and is commonly grown in a wild, semi-wild, and cultivated state in the land areas of the equatorial tropics. The fruit is the source of two types of vegetable oil, palm oil, and palm kernel oil, the better grades being used in the manufacture of margarine and compound cooking fats. The coconut tree is commonly grown in coastal areas of the tropics and sub-tropics. The bulk of the world's coconut crop is dried to produce copra from which the oil is later extracted. Sunflowers are an extremely adaptable species and can be grown in both temperate and tropical countries.They are probably most important in the USSR and Eastern Europe. The seed of the flax plant is called linseed. It is a subtropical and warm temperate crop not generally suited to the tropics. Major producers include the USA, Argentina, and the USSR. Linseed oil is a drying oil used principally in the manufacture of paints and varnishes rather than in the food industry.

The rapes and mustards produce oil-bearing seeds which can be grown in temperate northerly climates. The genus *Brassica* contains about 100 species, including the cabbages, turnips, swedes, oilseed rapes, and some mustards. Rape (*Brassica campestris* L. and *Brassica napus* L.) seeds yield an edible oil and, as with the other oilseeds, the residue can be used for feeding farm livestock. Major producers include India and China. The mustards include *Brassica juncea* (brown mustard) and *Sinapis alba* (white mustard) and the seeds, as well as being a source of condiment mustard, are utilized as a source of edible oil.

1.4.1 Nitrogen fixation

A distinguishing feature from the cereals is the ability of legumes to utilize atmospheric nitrogen and make it available to the host plant (Fig. 1.3). One of the effects of this is to produce seeds of high protein content which make them particularly valuable in the nutrition of man and domestic animals. This process is mediated by a symbiotic association of bacteria with the roots of the plant. The bacteria, *Rhizobia* spp., which are normally free-living in the soil, enter the legume roots from the seedling stage onwards. Subsequently, the host cells divide rapidly and characteristic nodules appear on the surface of the roots (Fig. 1.11). At the same time the bacteria enlarge, giving rise to the bacteroids and a characteristic red pigment, leghaemoglobin. Only after these events is nitrogen fixation possible. These bacteria, when free-living in the soil, do not fix nitrogen. There are a number of strains of *Rhizobia* spp. each with a broad range of specificity.

1.5 GROWTH CHARACTERISTICS OF THE SEED PLANTS

1.5.1 Cereals

The structural and physiological changes accompanying plant growth and development in the cereals are very similar, and, with the exception of maize this similarity extends into ear and seed formation.

The first sign in the field of successful germination is the appearance of the coleoptile and first leaf. The coleoptile is a transparent membrane surrounding the first leaf and disintegrates later in development. The most important feature of the early stages is tillering, which is the normal process of branch formation in cereals. Each tiller may result in the formation of an ear. The typical average is around three to four tillers per plant under normal cropping densities. By the time six leaves are unfolded on the main stem in barley, the potential number of grain sites or spikelets on the developing ear, hidden within the plant, have been determined. This stage is later in wheat.

The next stage of growth is characterized by stem elongation followed by heading or earing. This is a period of rapid growth during which there is competition for nutrients among the various growing parts of the plant. In particular, the last-formed spikelets may abort on the ear and the youngest tillers die off. Nitrogen can be applied at this time to maximize dry matter production and tiller survival. A number of the subsequent stages of growth are estimated by reference to the appearance of nodes on the main stem. These are thickened regions, or joints, whose function is partly in the distribution and transport of nutrients into the leaves and ears. The stem then grows by elongation between the nodes. By the time the second node can be detected in wheat all of the spikelets have been formed. As growth proceeds the ears push up inside the leaf sheaths surrounding the stems of the main shoot, and the flag leaf, or last leaf on the stem or tiller, appears. Subsequently the flag leaf sheath swells as the developing ear expands. In barley, fertilization takes place before the ear emerges, i.e. when only 2–3 cm of awns are protruding from the top of the flag leaf sheath. In wheat, the ear emerges, its base is about 5 cm above the flag leaf, and the anthers are extruded before pollen is shed and fertilization takes place. Following fertilization, the endosperm and the embryo grow rapidly, and together with the modified ovary wall they form the grain. Final grain dry weights for barley and wheat are between 30 and 50 mg. The general habit of a mature wheat plant is shown in Fig. 1.4. The growth habit of maize (American corn) is different from that of barley and wheat (Fig. 1.5). It is more robust and much larger. The grain sorghums have a similar general appearance.

The maize plant grown for grain production is normally a non-tillering type and, in general, single-eared varieties are preferred. The mature plant can grow to a height of between 4 and 5 m and is supported by a central woody stem. Maize differs from the other important cereal grasses of the world mainly in the nature of its inflorescences or flowering parts. The male

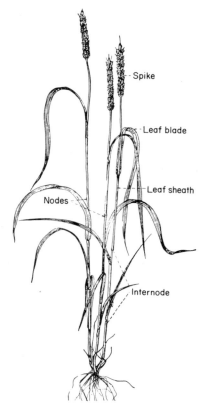

Figure 1.4   Full-grown wheat plant (Robbins, 1931)

flowers are located in spikelets which make up tassels at the top of the plant and the female flowers are in pairs on the cob or ear. Silks (styles) are produced by the female flowers a few days after pollen shedding and these continue to elongate until fertilized by the pollen. The surface of the silks is hairy and mucilaginous and thus facilitates the trapping of the pollen grains. After the pollen grains germinate, the pollen tubes penetrate the tissues of the silks and grow towards the embryo sacs where fertilization takes place. The endosperm soon begins to accumulate starch and other storage materials (Chapter 2), and by 50–60 days after pollination the kernel is ripe and ready for harvesting. As the ovary wall becomes modified to form the outer covering or pericarp of the maize grain it does not, in contrast to most other cereal grains, develop any green photosynthetic tissues. The entire ear is enclosed in modified leaf sheaths or husks.

1.5.2 Food legumes or pulses

One of the most familiar legumes is the common bean (*Phaseolus vulgaris*). The species contains a number of different varieties, which may be divided

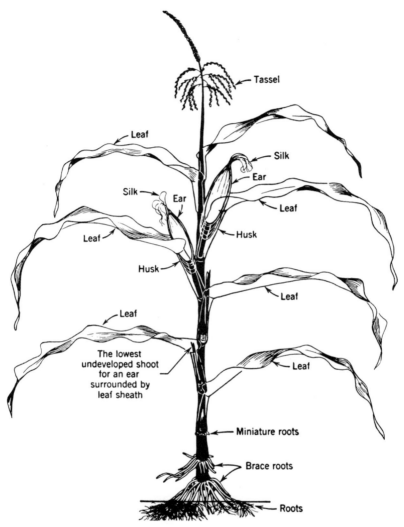

Figure 1.5 Full-grown maize plant (Wallace and Bressman, 1949)

into two categories: the dwarf or bush varieties, which do not require support and mature early, and the climbing varieties, which require support, take longer to mature, and have a longer bearing season.

The events following germination are similar in all varieties. After water has been taken up by the seed, the coat bursts and the root starts downwards growth into the soil (Fig. 1.6). The hypocotyl then elongates and bends, forming a hook which is forced upwards through the soil. The cotyledons and shoot apex are then raised above the ground by a gradual straightening of the hypocotyl.

The pronounced tap root grows rapidly to a depth of about 1 m. Extensive

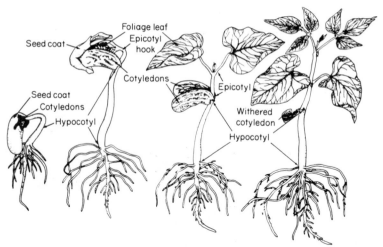

Figure 1.6   Stages in the germination of a bean (*Phaseolus vulgaris*) seed (Weier, *et al.*, 1974). (Reproduced by permission of John Wiley and Sons)

lateral roots are confined mainly to the top layer of soil. In the bush varieties the main stem of about 20–60 cm tall has between 4 and 8 nodes and terminates in an inflorescence. The climbing varieties might be 2–3 m tall with around 11–30 elongated nodes. The flowers are few in number and in a variety of colours from white to violet–purple. The plants are self-fertilized and pollination takes place at the same time as the flower opens. The flowers have a single ovary which contains several ovules. After fertilization, the ovary wall or pericarp elongates and expands rapidly, and the ovules develop into mature bean seeds. The pericarp is green throughout much of seed maturation—as are the seeds themselves—and is called the pod or legume. As with the cereals, a proportion of the ovules fail to develop into mature seeds. At maturity the pods split open along the midrib and the seeds are subsequently dispersed (Fig. 1.7).

The growth and development of the garden pea plant have much in common with those of the common bean. It is a short-lived annual plant with dwarf, semi-dwarf, and tall varieties, most of which have weak stems requiring support. The flowers are almost entirely self-fertilized and pods are generally green, each containing from 2 to 10 seeds (Fig. 1.8). Although the overall pattern of seed development is broadly similar in the pulses, considerable variation in plant growth habit does exist. For example, the lentil plant is an erect, branched annual about 25–40 cm tall. It matures in about 31 months, producing smooth oblong pods rarely more than 1.3 cm long and containing 1–2 seeds. Bengal gram (chick pea) is probably the most important pulse in India and is an erect or spreading much-branched annual, about 25–50 cm tall. It is harvested 4–6 months after sowing and the swollen oblong pods are 2.3 × 1.2 cm and contain 1–2 seeds (Fig. 1.9). Like Bengal gram, black gram is widely cultivated in India. It is an erect very hairy

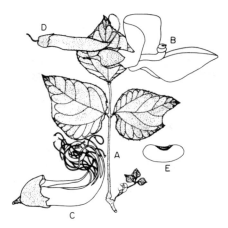

Figure 1.7 The common bean (*Phaseolus vulgaris*). A, Leaf ($\times \frac{1}{4}$); B, Flower ($\times 1\frac{1}{2}$); C, flower with corolla removed ($\times 2$); D, young pod ($\times 1\frac{1}{2}$); E, seed ($\times 1$) (Purseglove, Vol. 1, 1968). (Reproduced by permission of the Longman Group Ltd)

Figure 1.8 The garden pea (*Pisum sativum*). A, Flowering shoot ($\times \frac{1}{5}$); B, flower ($\times \frac{3}{5}$); C, flower in longitudinal section ($\times 1$); D, young pod ($\times 1$); E, young pod opened ($\times 1$) (Purseglove, Vol. 1, 1968). (Reproduced by permission of the Longman Group Ltd)

plant varying in height from 30 to 90 cm. The pods are brown and very hairy, about 4–7 cm long at maturity, and contain 6–10 seeds. Seeds are oblong with square ends, about 4 mm long, and are generally black in colour. The pigeon pea is grown throughout the tropics and subtropics with regional production greatest in the Far East. The plant is a woody, short-lived perennial shrub, 1.4 m in height. As a pulse crop, it is probably best grown as an annual. Growth is slow and the late varieties can take as long as 9–12 months to reach

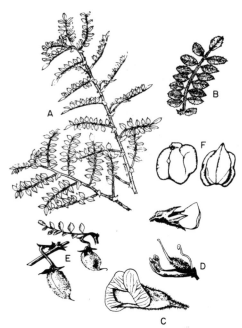

Figure 1.9 Chick pea (*Cicer arietinum*). A, Leafy shoots ($\times \frac{1}{5}$); B, leaf ($\times \frac{3}{5}$); C, flower ($\times 2$); D, stamens and pistil ($\times 2$); E, pods ($\times \frac{1}{5}$); F, seeds ($\times 1$) (Purseglove, Vol. 1, 1968). (Reproduced by permission of the Longman Group Ltd)

maturity. Flowering extends over several months. The pods are about 4–10 cm in length, generally hairy, and green or maroon in colour with 2–8 seeds in each. The seeds are generally round or oval, about 8 mm in diameter.

1.5.3 Oilseed legumes

Many soybean varieties are grown in the Far East. These vary in time to maturity, height, and shape, as well as colour, size, oil and protein content, and quality of the mature seeds. Most plants are erect, bushy annuals with a long tap root, and vary in height between 20 and 180 cm. The flowers open early in the morning and the pollen is shed, just before or at the time of opening, directly on to the stigma. The plants are normally self-pollinated. The resulting pods are borne in clusters on short stalks in groups of 3 to 15. They are hairy, vary in colour from pale yellow to black, and usually contain 2–3 seeds (Fig. 1.10).

The groundnut or peanut is a low-growing annual plant which may be erect or trailing. The flowers develop either singly or in clusters of three and are most numerous towards the base of the plant (Fig. 1.11). The first flowers appear 4–6 weeks after planting and the maximum number is produced after a further 4–6 weeks. They are generally self-pollinated. Immediately after

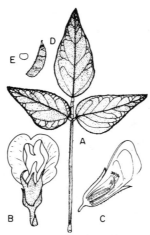

Figure 1.10 Soybean (*Glycine max*). A, Leaf (×¹/₄); B, flower from below (×4); C, flower in longitudinal section (×4); D, pod (×¹/₄); E, seed (×¹/₄) (Purseglove, Vol. 1, 1968). (Reproduced by permission of the Longman Group Ltd)

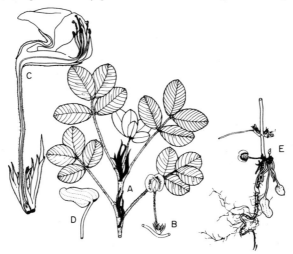

Figure 1.11 Groundnut (*Arachis hypogaea*). A, Leafy shoot (×¹/₄); B, flower (×¹/₄); C, flower in longitudinal section (×4); D, developing fruit (×¹/₄); E, base of plant showing flower, developing fruits, and roots with nodules (×¹/₄) (Purseglove, Vol. 1, 1968). (Reproduced by permission of the Longman Group Ltd)

fertilization the ovary is pushed downwards into the soil, as a result of rapid cell division and elongation of tissue at its base. Normally the ovary begins to develop only when it is 2–5 cm into the soil, where it gradually attains a horizontal position with the ovules parallel to the soil surface. The mature fruit usually contains two seeds and is externally constricted between them. The pod wall is thick and consists of an outer spongy layer, a middle fibrous and woody layer, and an innermost layer which lines the pod. The layer serves as storage tissue during development and collapses as the pod matures, eventually forming a thin papery lining.

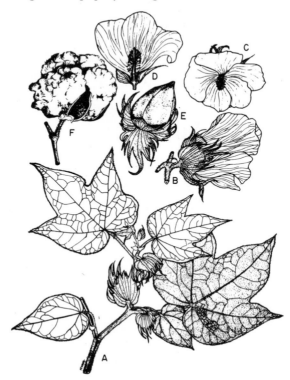

Figure 1.12 Upland cotton (*Gossypium hirsutum* var. *latifolium*). A, Sympodial branch (×¹/₄); B, flower from side (×¹/₃); C, flower from above (×¹/₃); D, flower in longitudinal section (×¹/₃); E, unopened boll (×¹/₃); F, opened boll (×¹/₃) (Purseglove, Vol. 2, 1968). (Reproduced by permission of the Longman Group Ltd)

1.5.4. Non-leguminous oilseeds

The cotton plant is an annual warm-season plant which requires a frost-free growing season of about 6–7 months. Upland cotton (*Gossypium hirsutun* var. *latifolium*) is a subshrub, 1–1.5 m tall (Fig. 1.12). The first flowers are produced 8–10 weeks after planting and flowering normally continues for 2 months. The seeds are contained within spherical or ovoid leathery capsules, called bolls. These grow to full size in about 25 days after opening of the flowers and the seeds develop for a further 25 days before the boll opens. The mature boll contains several valves or 'locks', each of which can contain up to 9 seeds. Two types of unicellular hairs are attached to the seed epidermis, short hairs or fuzz and long, easily removable white hairs or lint which are the cotton fibres of commercial practice (Fig. 1.23B). The seeds are separated from the lint in a process called ginning and are used mainly in the production of seed oils and protein-rich cotton seed-cakes or meals (Section 4.3).

The oil palm (Fig. 1.13) is a stout, single-stemmed, upright growing palm, growing to a mature height of around 21 m and carrying a crown of 20–25 pinnate leaves. It is a sun-loving plant which grows best at a mean annual temperature of between 24 and 27 °C. Germination is extremely slow, and seeds usually require some form of heat treatment before they can be induced to germinate. Male and female flowers occur separately on the same plant although, when the flowers reach sexual maturity, most are found to be female. About 3 years after flower initiation the flowers open and most of the pollen is shed within the next 2–3 days. The plants are almost exclusively wind-pollinated. Ripening of the fruit takes place from $4^1/_2$ to 6 months after pollination. Although cultivated oil palms begin to fruit at about 4 years after planting, full production is obtained only after 12–15 years. The plants continue to bear fruit for 40–50 years. The fruits are borne tightly clustered in large bunches. They vary in shape from nearly spherical to ovoid or elongated, bulging slightly at the top, and between 2 and 5 cm in length. The fruit consists of a fairly thin outer skin or exocarp, a pulpy mesocarp, and a hard, thin, stony endocarp which surrounds the single seed or kernel. The outer pulp or soft part of the fruit is the source of palm oil and the seed or kernel is the source of palm kernel oil.

Figure 1.13　An oil palm (*Elais guineensis*) (Acland, 1973). (Reproduced by permission of the Food and Agriculture Organization of the United Nations)

The coconut tree (*Cocos nucifera* L.) is an unbranched palm growing to a mature height of around 24 m and topped by a crown of large pinnate leaves. The trees require a warm sunny climate with an ideal mean annual temperature of about 29 °C.

Germination is slow and it can be 4 months after planting the ripe nut that shoots first appear. The young palms are usually transplanted to the field after about 30 weeks, by which time the first 3 seed leaves have developed.

Tall coconut palms (var. *typica*) begin to flower and bear fruit after 6 or 7 years' growth.

Like the oil palm, male and female flowers occur separately on the same plant and are born on a branched structure called the spadix (Fig. 1.14). The female flowers lie at the base of the branches and the male flowers, of which there might be ten times as many, lie on the upper parts of each branch. Since it is rare, on any one tree, for the pollen to be shed at the same time as the female flowers open, cross-pollination is commonly observed. The colour of the fruit, initially a bright shining green, changes to yellow with the onset of maturation. The endosperm begins to develop when the fruit is about 6 months old and at first forms a greyish translucent jelly surrounding a large cavity, which may contain up to half a litre of 'coconut water'. The endosperm subsequently develops into a hard white layer and only in the later stages of fruit maturation does it begin to accumulate the economically important oil reserves. The fruits are ripe about 12–14 months after pollination. The trees reach full bearing only after 10 or 12 years but they produce fruit continually through the year and throughout a lifetime of some 60 or more years. The fibrous fruit (Fig. 1.25A) is extremely large, up to 30 cm in length, oval in shape, and weighing as much as 2 kg. The exocarp or outer husk is thick and fibrous and surrounds a hard bony endocarp or shell. The single seed lies within the shell, and is closely attached to it via the testa.

0.5 m

Figure 1.14 A coconut inflorescence. Only five female flowers can be seen (Acland, 1973). (Reproduced by permission of the Food and Agriculture Organization of the United Nations)

The rapes, which resemble both swedes and turnips, are biennial in habit. During the first year, a swollen fleshy 'root' is formed and in the second year the food reserves of this organ are used for the formation of the flowers, fruits, and seeds. The stem is erect with large green leaves and is from 60–120 cm in height. The flowers are bright yellow, borne on long, slender stems. The fruit is a siliqua, 4–7 cm in length, which contains the small (2 mm diameter) oval, reddish to black seeds.

White mustard is a hairy annual herb, about 30–80 cm high. The pods are 2.5–4 cm in length and the seeds are pale yellow to white in colour, more or less spherical, and over 2 mm in diameter.

1.5.5 Coffee

A number of species of the genus *Coffea* are known, the most important being *Coffea arabica* L. (Fig. 1.15), which produces most of the world's coffee. Ideal conditions for growth are found on the equator at approximately

Figure 1.15 Arabica coffee (*Coffea arabica*). A, Shoot ($\times\,^3/_{10}$); B, portion of under-surface of leaf ($\times\,^3/_{10}$); C, portion of shoot with star flowers ($\times\,^3/_{10}$); D, star flower ($\times\,^3/_{10}$); E, normal flower in longitudinal section ($\times\,^3/_{10}$); F, fruiting node with some of the fruits removed ($\times\,^3/_{10}$); G, fruit ($\times\,^3/_{10}$); H, fruit with part of mesocarp removed ($\times\,^3/_{10}$) (Purseglove, Vol. 2, 1968). (Reproduced by permission of the Longman Group Ltd)

1 500–1 800 m and temperatures between 15 and 24 °C. The plants are evergreen glossy-leaved bushes, which can grow to 5 m if unpruned. The viability of the seeds is comparatively short and they should be planted within 2 months, or even less, after harvesting. The cotyledons emerge between 3 and 8 weeks after planting, and the young seedlings are generally raised in nurseries before transplanting in the field at about 6–10 months old. The trees come into bearing 3–4 years after planting and the fragrant white flowers are borne in clusters. Under adverse conditions, particularly at high temperatures, abnormal or star-flowers are produced. These have no functional stamens and fail to set fruits. The endosperm starts growth about 21–27 days after pollination, normally self-pollination, and by 60–70 days the embryo begins to develop. By $3^1/_2$ months the embryo is shield-shaped and easily recognizable. Maturation takes between 7 and 9 months. The fruit, which changes in colour from green to yellow and finally crimson over the maturation period, is about $1^1/_2$ cm in length when mature. It consists of a smooth, tough outerskin or exocarp, a soft yellowish pulp or mesocarp and a greyish, fibrous endocarp or parchment which surrounds the seeds. There are normally two seeds per fruit.

The rest of world coffee production is met by the species *Coffea canephora* or Robusta coffee. This variety, which originated in Africa, is now widely distributed throughout the tropics and can be grown at lower elevations unsuited to *C. arabica*.

1.5.6 Cocoa

The cocoa bean is the seed of the tropical tree *Theobroma cacao* (Fig. 1.16), which appears to have come originally from the rain forests of the upper

Figure 1.16   Cocoa tree (*Theobroma cacao*). Drawing of a cocoa tree bearing six ripe pods. The tree has two chupons and two sets of fan branches

Amazon but is now cultivated in many tropical countries, including West Africa and New Guinea. *T. cacao* requires humid sheltered conditions for growth and is normally restricted to land below 1 000 m and within 10° of the equator. It rarely reaches more than 8 m in height and has a very characteristic growth pattern and requirements. The trees are propagated from seeds or cuttings which are planted out at distances determined by the soil quality and tradition but generally not less than 2 m. During the early years whilst the trees are establishing themselves, shade is essential, and for this reason, when a cocoa plantation is prepared, a proportion of the original forest trees is left. Furthermore, a low ground cover is often provided in the first few years by planting food crops such as cassava or bananas. As growth of the young plant proceeds it sends a tap root straight down to a depth of 2 m in good soil and a single main stem or 'chupon' directly upwards to a height of 1–1.5 m. This point is reached when the trees are just over 1 year old. The terminal bud then sub-divides and further growth results in a fan of 3–5 branches arising from a single point (the jorquette). Further growth occurs when a second chupon arises from just below the jorquette and again results in a second, higher fan. This process may repeat itself several times but the actual appearance of trees in plantations depends more on the local technique of pruning than on the natural tendency of the tree. If cuttings are taken, chupons produce chupons and fan branches fans, although chupons often arise from a fan branch later.

The trees normally begin to flower after 3–4 years and peak yields occur between 8 and 10 years. The flowers are produced on the older, leafless parts of the trunk and branches in great numbers, but as the fertilization rate is low only relatively few fruits are produced, perhaps as low as one fruit for every 500 flowers. As the flowers appear ill-adapted to self-pollination, have no scent or nectar, and the pollen is too sticky for wind pollination, the mechanism of fertilization remained a mystery for a long time but it is now believed that ceratopogonid midges are mainly responsible. These midges feed on the purple tissue of the staminodes and guide-lines of the petal pouches.

After pollination, the fruits develop and become ripe in 5–6 months. The seeds have no period of dormancy and remain viable for only a short time, so, as viability is essential in the first stage of cocoa production (Section 5.8) the fruits must be harvested and processed immediately on becoming ripe. This means that pods have to be assessed individually each day and the harvest period can extend over about 3 months. It is common for there to be two harvests a year: a major harvest in the October–February dry season and a minor harvest in the first few months of the wet season.

1.6 THE STRUCTURE OF MATURE SEEDS

The fruit is the fertilized and matured ovary of a flower. The ripened ovule containing the embryo is called a seed. In most seeds, integuments of the ovule become the hard coats of the mature seed. Differences in the structure of economically important seeds derive mainly from variation in the proportions of embryo and endosperm present.

All seeds have an endosperm derived from the initial triploid nuclear fusion within the ovule (Section 2.1). This tissue persists as a food storage tissue in relatively few dicotyledonous seeds, one exception being the coffee seed (Section 1.6.3), but is highly developed in cereals where it is of major economic importance. The embryo is a major part of the mature dicotyledonous seed and contains the principal food reserves. The cereal embryo, in contrast, comprises only a minor part by weight of the mature grain.

1.6.1 Cereal grains

These are one-seeded dry indehiscent fruits with the pericarp firmly attached to the seed. On the whole, the structure of the grains is remarkably similar and differences are generally those of shape, colour, and the nature of the seed coat. The structure of a mature wheat grain is shown in Fig. 1.17.

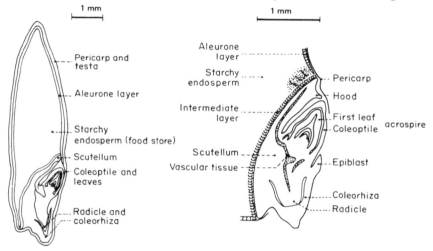

L.S. Wheat grain (*Triticum aestivum*)   L.S. Embryo

Figure 1.17   Structure of a mature wheat grain and embryo (Bewley and Black, Vol. 1, 1978). L.S. wheat grain; L.S. embryo

The grain is plump, 'white' or 'red' in colour, and generally referred to as naked, since, unlike barley, rice, and oats, it loses the outer layers derived from the flower during threshing. The dorsal surface of the grain is smooth and rounded, while the opposite or ventral side has a characteristic furrow called the crease. The apex is covered by a number of short, stiff hairs called the brush. The pericarp of the ripe grain is about 45–50 μm thick and consists at maturity of four or five layers of cells (Fig. 1.18), one of which is a

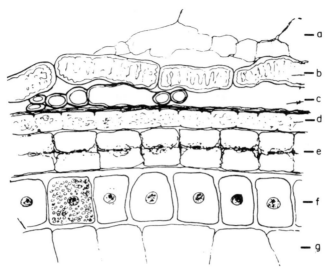

Figure 1.18 Transverse section through a mature wheat grain illustrating arrangement of outer layers. a, Transparent layer of pericarp; b, green layer of pericarp cross cells; c, tube cells; d, testa; e, nucellar epidermis; f, aleurone; g, endosperm (modified from Percival, 1921)

well defined layer of cells (cross cells) elongated in a plane at right-angles to the axis of the grain, and containing chlorophyll during the early stages of development. By maturity, these cells are crumpled and crushed, often with small intercellular spaces. Inside the cross cells is the inner epidermis of the pericarp, made up of widely spaced thin-walled cells known as tube cells. These in turn cover the testa, to which the colour of the grain is mainly due, which contains two layers of elongated cells originally derived from the inner tissues of the ovule wall. The nucellar layer lies beneath the testa and is a layer of cells with very little apparent structure. The interior of the grain is mainly occupied by the endosperm, which is a mass of cells containing starch granules and protein bodies. The outermost layer of the endosperm is made up of rectangular cells called the aleurone, and is about 65–70 μm across. The embryo is at the base of the grain on the dorsal side and is a highly differentiated organ. The shoot apex and the several rudimentary leaves are surrounded by a sheath called the coleoptile. Similarly, the root apex is surrounded by a sheath, the coleorhiza. Between the embryo and the endosperm is a fleshy shield-like structure, the scutellum.

The barley grain differs in a number of respects from the wheat grain. Firstly, the ovary wall fuses intimately with the inner surface of the palea, giving rise to a covered grain. Secondly, the aleurone layer in barley is several layers thick, except in the vicinity of the embryo, where it is reduced to a single layer. This proliferation of aleurone cells in barley is thought to account for its use in the brewing industry, since these cells are the source of starch-degrading enzymes synthesized during malting. The barley grain is

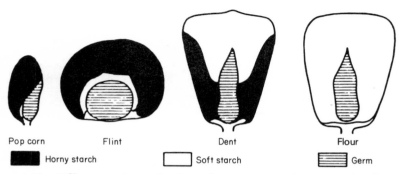

Pop corn   Flint   Dent   Flour

■ Horny starch   □ Soft starch   ▤ Germ

Figure 1.19 Differences in maize kernel structure and texture (Wallace and Bressman, 1949)

also less plump than the wheat grain and, because of the thick husk, the crease is not so obvious.

Maize has a wide range of varieties and this is demonstrated by a number of different kernel types. These include dent, flint, sweet, flour, and popcorn (Fig. 1.19). Dent, which is the most widely grown type of corn in the USA, is characterized by a depression or 'dent' in the crown of the seed. This is caused by a deposit of soft starch at the crown. The more soft starch in the deposit, the greater is the indentation. Flint corns differ from dent in that they contain less soft starch and it is more centrally located. The grains are rounded in cross-section and there is no 'dent'. Sweet corns are picked when immature and are characterized by kernels with so much sugar and so little starch that they are wrinkled when dry. Flour corn kernels are composed largely of soft starch and have little or no dent. Popcorn is generally characterized by small hard kernels and is really an extreme form of flint with the endosperm containing only a small proportion of soft starch. The general structure of a maize kernel is shown in Fig. 1.20. Like wheat it is a naked grain with the pericarp as the outer layer. Most maize kernels are yellow, but white varieties exist.

Oats and rice grains are substantially similar to barley but differ in that they have a single layer of aleurone cells and are extensively husked.

1.6.2 Legumes and oilseeds

Although the seeds of leguminous plants are generally similar in structure, there are significant differences in size and colour. The pod, or pericarp, contains the seeds and, unlike the cereals, the two can easily be separated. The outermost layer of the seed is the testa or seed coat. In most legumes the endosperm is short lived (Section 2.1.2), and by maturity is reduced to a thin layer surrounding the cotyledons (Fig. 2.4), or embryo. A typical example of a legume seed is probably that of the common bean (*Phaseolus vulgaris*). It is characterized by a number of external structures. These include the hilum, micropyle and raphe (Fig. 1.21). The hilum is a large oval scar near the middle

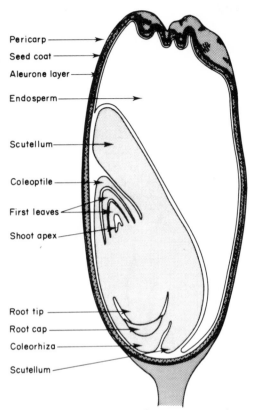

Pericarp

Seed coat

Aleurone layer

Endosperm

Scutellum

Coleoptile

First leaves

Shoot apex

Root tip

Root cap

Coleorhiza

Scutellum

Figure 1.20   Median longitudinal section of mature maize kernel. (Weier *et al.*, 1974).
(Reproduced by permission of John Wiley and Sons)

of one edge, left where the seed broke away from the stalk. The micropyle is a small opening in the seed coat beside the hilum and was originally the site where the pollen tube entered the ovule. The raphe is a ridge at the side of the hilum opposite the micropyle and represents the base of the stalk which by maturity has fused with the seed coat. When the seed coat of a bean is removed, for example after soaking, the endosperm comes off with it and the remainder is composed entirely of embryonic structures. These include the shoot, which consists of two fleshy cotyledons, a short axis below the cotyledons, and above the cotyledons a short axis which has several foliage leaves and terminates in the shoot tip.

The soybean seed is ovoid or nearly spherical and up to 12 mm in length. The hilum is distinct with a clear slit and the embryo shows the usual leguminous feature of two bulky cotyledons. Like most leguminous seeds, the testa epidermis consists of elongated or palisade cells (Fig. 1.22A), and is many times thicker than the layer of endosperm beneath. In contrast to some of the legumes, the soybean endosperm has a distinct aleurone layer with oil droplets and small aleurone grains.

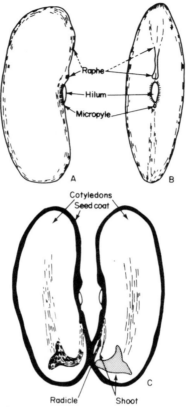

Figure 1.21   Bean (*Phaseolus vulgaris*) seed. A, External side view; B, external face or edge view; C, embryo opened (Weier *et al.*, 1974). (Reproduced by permission of John Wiley and Sons)

Groundnut seeds vary in shape from almost spherical to roughly cylindrical and in length from 10 to 20 mm. The embryo consists of two elongated cream cotyledons. The testa has no palisade cells (Fig. 1.22B), but consists of cells where outer walls are thickened and pitted. The rest of the testa grades from a compact layer of cells immediately underneath the outer epidermis to spongy layers where the cells become more squashed. The endosperm is a single layer of cells containing oil droplets and possible aleurone grains. The embryo consists of cells with characteristically pitted walls.

Cotton seeds are on the average about 10×6 mm in size and contain highly convoluted cotyledons surrounded in turn by a thin papery endosperm, a few nucellar fragments, and a seed coat consisting of about five cell layers (Fig. 1.23A). The nucellus and endosperm adhere to the embryo and not to the husk when the seed is broken. Resin glands are often conspicuous within the embryo cells and contain a greenish black opaque secretion. These are the sites of secretion of the toxic polyphenol gossypol (Section 4.3.4.3).

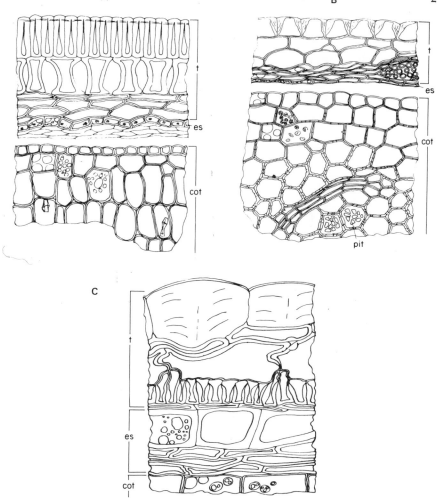

A B C

Figure 1.22 Transverse sections of soybean (A) (×230); groundnut (B) (×180); and mustard (C) (×380); seeds, showing testa (t), endosperm (es) and cotyledon (cot) (Vaughan, 1970)

The oil palm seed or kernel lies within a closely adhering fibrous hard outer shell, the whole generally being referred to as a nut (Fig. 1.24). The proportion of the nut present as shell varies greatly. For example, nuts from the variety *dura* are about 20–40% shell, whereas those from the variety *tenera* are only 5–20% shell. Typical *dura* nuts are 2 to 3 cm in length, *tenera* nuts usually 2 cm or less. The kernel consists of layers of hard oily endosperm, greyish white in colour and surrounded by a dark brown testa covered with a network of fibres. The embryo, which is about 3 mm in length, is embedded in the endosperm.

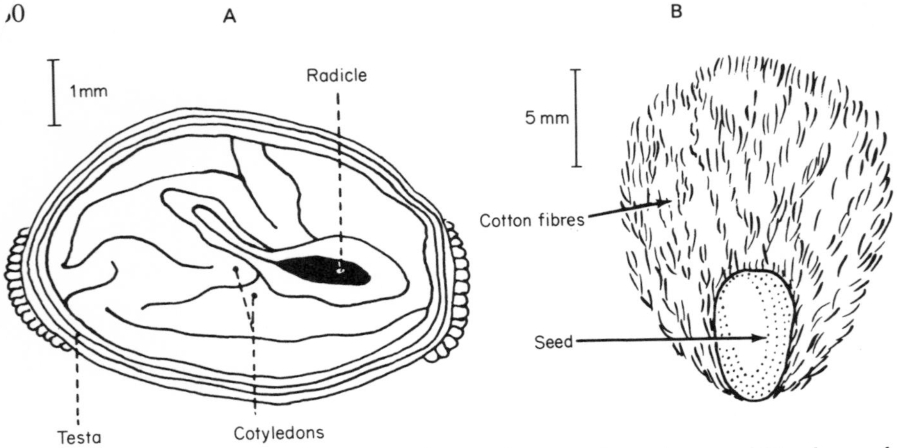

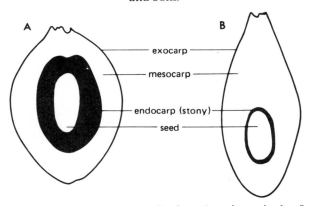

Figure 1.23   Cotton seed. A, Cross-section of *Gossypium barbadense* seed (Bewley and Black, 1978). (Reproduced by permission of Springer-Verlag.) B, cotton seed with attached cotton fibres (Weier *et al.*, 1974). (Reproduced by permission of John Wiley and Sons)

Figure 1.24   Oil palm fruit. A, Longitudinal section through the fruit of a thick shelled type ($\times 1$); B, longitudinal section through the fruit of a thin-shelled type ($\times 1$) (Cobley, 1976). (Reproduced by permission of the Longman Group Ltd)

    The coconut seed (Fig. 1.25A) lies within a hard outer shell or exocarp. The nut, which includes both shell and seed, is 10–15 cm in diameter at maturity and can weigh as much as 1.5 kg. The seed itself consists of a hard, white, oily endosperm layer which surrounds the central cavity. The endosperm layer itself is surrounded on the outside by a brown, paper-thin testa closely attached to the outer shell. The small embryo is located at the base of the fruit.

    In rape and mustard seeds—and the Brassicas in general—seed structure more or less corresponds. The seed coat or testa of *Sinapis alba* (white mustard) consists of 4–5 layers of cells including an outer layer of larger cells with finely beaded walls, and an inner palisade layer (Fig. 1.22C). Inside this is the endosperm, which consists of an outer layer of cells, with thickened colourless walls, containing fat globules and small aleurone grains, and an inner, very narrow, layer of compressed cells without any evident cellular structure. The centre of the seed is filled by the cotyledons.

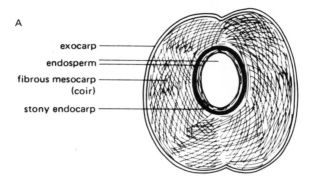

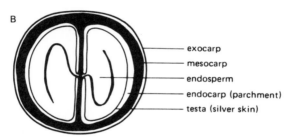

Figure 1.25 A, Coconut (*Cocos nucifera*), diagrammatic section through a fruit ($\times \frac{1}{6}$); B, Arabica coffee (*Coffea arabica*), diagrammatic section through the fruit ($\times 2$) (Cobley, 1976). (Reproduced by permission of the Longman Group Ltd)

1.6.3 Coffee beans

Coffee seeds are between 8.5 and 12.5 mm in length. Inside the fruit (Fig. 1.25B) the two seeds are pressed together, resulting in each having one flattened surface with a crease down the middle. The rest of the seed is convex in shape. Directly under the parchment layer or endocarp lies the thin silvery testa, or silver skin, which forms the outer surface of the seeds. The seeds themselves consist mainly of a horny green endosperm and a small embryo near the base. The dried seeds, after removal of the silver skin, are the coffee beans, which, after roasting and grinding are used in the making of coffee (Section 5.12).

1.6.4 Cocoa beans

The cocoa fruit (Fig. 1.26) is usually referred to as a pod, although botanically it is a drupe or berry. Pods are cylindrical to nearly spherical in shape and 10–30 cm long. The exact shape, surface texture, and colour vary considerably (see later comments on nomenclature). Each pod has a thick, fleshy pericarp or husk associated with a partially lignified mesocarp. In the centre of the pod are five rows of seeds or beans, usually between 20 and 60 in number. The beans are surrounded by a watery, sugary, mucilaginous pulp which is derived from the outer integument of the ovule.

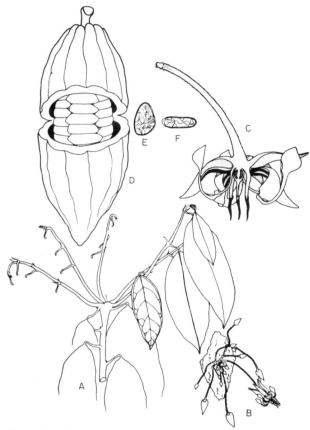

Figure 1.26   Cocoa (*Theobroma cacao*). A, Jorquette with young fan branches ($\times \frac{1}{7}$); B, cauliflorus inflorescence ($\times \frac{3}{5}$); C, flower in longitudinal section ($\times 2\frac{1}{2}$); D, fruit ($\times \frac{3}{10}$); E, seed in longitudinal section ($\times \frac{3}{10}$); F, seed in transverse section ($\times \frac{3}{10}$) (Purseglove, Vol. 2, 1968). (Reproduced by permission of the Longman Group Ltd)

The beans are variable in size but are usually ovoids in the size range 2–4 × 1.2–2 cm. Within a leathery skin or testa, the beans consist largely of two large, fatty cotyledons with a small embryonic axis and the thin membranous remains of the endosperm. On sectioning, the colour of the cotyledons can vary from white to deep purple. An unusual feature of the mature cocoa bean is a relatively high water content, 35–50%. This has to be maintained if the seeds are to retain their viability for even a few weeks.

There is some confusion about where the use of the botanical term 'cacao' should cease and the use of 'cocoa', the name of the eventual product, begin. The modern trend seems to be to restrict 'cacao' to strictly botanical use and to employ 'cocoa' for all other purposes. This practice has been followed in this book and so we use *Theobroma cacao* as the name of the tree but cocoa pods, cocoa beans, and cocoa butter as terms for the objects and materials which are used or formed in the preparation of cocoa (Section 5.8).

Difficulties also exist in naming the many varieties of *T. cacao* which are grown. Botanically, two subspecies and a number of forms have been recognized, but in the commercial world three main types appear to be recognized, based on the original Venezuelan categories. These are:

(a) Criollo (or native) where the ripe pods are yellow or red with deep furrows and a conspicuous point. The seeds are large and round and the colour of the cotyledons is white or pale violet. These beans give low yields of high-quality cocoa lacking any astringency.

(b) Forastero (or foreign). This term refers to trees similar to the Brazilian trees first introduced into Venezuela hundreds of years ago. The ripe pods are yellow, relatively spherical and smooth, and the fresh cotyledons are dark violet in colour. Although the cocoa produced from this type may not be of the best quality, the trees grow well and have a high yield. For this reason Forastero types have been planted in many countries including West Africa (the Amelando form) and Ecuador (Cacao Nacional).

(c) Trinitario. Pod and seed appearance varies considerably in the wide range of Trinitario types, which are hybrids between Criollo and Forastero typically occurring in Trinidad. Many of these strains are vigorous and high-yielding in comparison with Criollo but are still capable of giving a 'fine' cocoa. Trinitario types are used extensively in breeding programmes.

REFERENCES

General references

Briggs, D. E. (1978). *Barley*, Chapman and Hall, London.

Cook, A. H. (1962). *Barley and Malt*, Academic Press, New York and London.

FAO (1961, Reprinted 1978). *Agricultural and Horticultural Seeds—Their Production, Control and Distribution*, FAO, Rome.

Godin, V. J. and Spensley, P. C. (1971). *Oils and Oilseeds*, Tropical Products Institute, London.

Hector, J. M. (1936). *Introduction to the Botany of Field Crops, Vol. 1, Cereals, Vol. 2, Non-Cereals,* Central News Agency, Johannesburg.

Mangelsdorf, P. C. (1974). *Corn: its Origin, Evolution and Improvement*, Harvard University Press.

Ray, P. M. (1972). *The Living Plant*, 2nd ed., Holt, Rinehart, and Winston, New York.

Specific references

Acland, J. D. (1971). *East African Crops*, Longmans, London.

Bewley, J. D. and Black, M. (1978). *Physiology and Biochemistry of Seeds*, Springer-Verlag, Berlin.

Cobley, L. S. (1976). *An Introduction to the Botany of Tropical Crops*, 2nd ed. revised by W. M. Steele. Longmans, London.

FAO (1977). *The Fourth World Food Survey*, Food and Nutrition Series No. 10, FAO, Rome.

Milner, M. (1975). *Nutritional Improvement of Food Legumes by Breeding*, Wiley, New York.

Percival, J. (1921). *The Wheat Plant*, Duckworth & Co., London.

Purseglove, J. W. (1968). *Tropical Crops, Dicotyledons*, Vols. 1 and 2, Longmans, London.

Robbins, W. W. (1931). *The Botany of Crop Plants*, P. Blakiston's Son & Co. Inc., Philadelphia.

Vaughan, J. G. (1970). *Structure and Utilisation of Oilseeds,* Chapman and Hall, London.

Wallace, H. A. and Bressman, E. N. (1949). *Corn and Corn Growing*, 5th ed., Wiley, New York.

Weier, T. E., Stocking, C. R., and Barbour, M. G. (1974). *Botany: An Introduction to Plant Biology*, 5th ed. Wiley, New York.

Chapter 2

Seed formation

2.1 MORPHOLOGICAL CHANGES DURING SEED FORMATION

The morphological and biochemical changes which accompany seed maturation are of particular interest since they lead to the identification of factors involved in the control of yield and composition of the harvested seeds.

The initial events following fertilization are similar in most flowering plants. The pollen germinates on the stigma forming a pollen tube, which penetrates the style and carries two male nuclei into the embryo sac. One nucleus fuses with the nucleus of the egg cell forming the zygote. The second male gamete nucleus enters the central cell and unites with the two polar nuclei to form the primary endosperm nucleus. The subsequent development of the zygote and endosperm differs considerably in different groups of plants. In cereals, for example, the endosperm grows rapidly to become the major storage organ of the mature grain. In the legumes, on the other hand, which include the oilseeds, soybean, and groundnut, as well as peas and beans, the embryo grows to fill the seed almost completely at maturity and the cotyledons become the major storage organs. Since the endosperm is generally a starchy tissue, and embryos are normally reservoirs of lipid and protein, the differences in nutritional composition between the cereals and legumes are easily explained.

2.1.1 Cereal grains

The grain maturation period varies between and within species and depends heavily on the climatic and environmental conditions during development. For example, barley might take only 40 days to reach maturity from fertilization in the Canadian prairies but as long as 70 days in Scotland.

After fertilization, the initial triploid endosperm nucleus rapidly divides and by about 2–3 days after fertilization there might be from 200–500 free endosperm nuclei within the embryo sac in barley. In wheat this figure might be 5 000. Cell wall formation commences about this stage, the exact timing depending upon species and environmental conditions. Starch deposition within amyloplasts commences early in development, as does the formation of small spherical and oval membrane bound structures which may be protein bodies. The rate of cell division in the embryo is less than in the endosperm and throughout development this tissue never accounts for much more than 3–5% of total seed dry weight. Embryo development is characterized by the appearance of protein bodies, amyloplasts, extensive endoplasmic reticulum, mitochondria, and large numbers of ribosomes. Starch is not a major embryo storage product.

Forming part of the endosperm, but distinct in structure, is the aleurone layer which is first distinguished early after anthesis. The cells are initially cuboidal in shape with thin walls and large nuclei. This layer is probably

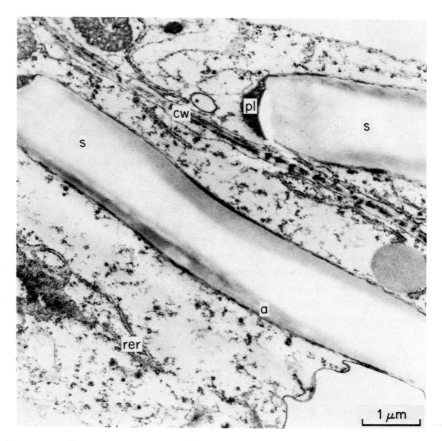

Figure 2.1 Electron micrograph of barley endosperm from a grain 23 days after anthesis showing: a, amyloplast; s, starch granule; cw, cell wall; pl, plastid lamellae; rer, rough endoplasmic reticulum. (Reproduced by permission of Dr. M. P. Cochrane)

developed to its greatest extent in barley where it is 3–4 layers thick by maturity.

Endosperm cell division largely ceases by around 3–4 weeks after anthesis and final cell numbers in wheat, for example, are between 100 000 and 150 000. Growth subsequently proceeds by means of cell expansion, the nuclei and mitochondria becoming obscured as the cells become filled with starch and protein. In general, the starch granules, which are contained within double membrane bound bodies called amyloplasts (Fig. 2.1) grow steadily throughout development, reaching a maximum diameter in wheat of around 25 μm and in normal maize of about 20 μm. In some cereals, for example barley and wheat, there is a secondary initiation of small granules (<10 μm) around 2 weeks after anthesis. These remain small. At maturity in most cereals there will be a range of granule sizes with a generally broad division into large and small. In wheat particularly, this division is into two types, as well as two sizes, since the large starch granules have a characteristic lenticular shape and are surrounded by a peripheral groove, and the smaller granules are spherical (Fig. 2.2). The protein deposits of endosperm and aleurone are distinct, those of the former often being called protein bodies and those of the latter aleurone grains. Both appear to be membrane bound organelles around 1–2 μm in diameter. In maize endosperm they have been shown to be the site of zein deposition and are associated with components

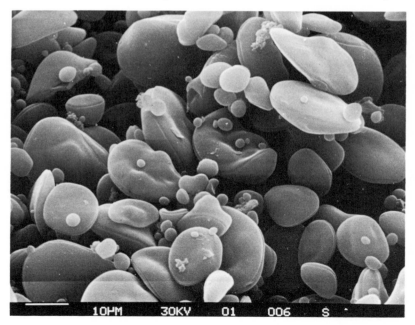

Figure 2.2   Scanning electron micrographs of starch granules from immature wheat endosperm. Large granules have a characteristic lenticular shape with a peripheral groove

such as polyribosomes which are specific for the synthesis of zein. Thus protein accumulates steadily within increasing numbers of small granules as development proceeds. The aleurone grains are particularly interesting in that, unlike the protein bodies of endosperm, they contain inclusions and account for much of the phytic acid deposits characteristic of the outer layers of cereals. Lipid droplets also surround the outside of the aleurone grain. In spite of the phytic acid, which is known to reduce the availability of some essential mineral elements (Section 2.7.1), the aleurone is a highly nutritious part of the cereal grain and its removal results in a decrease in the nutritive value of the grain.

Complex structural changes also occur in the maternal tissues surrounding the endosperm and embryo during the developmental period. In barley, wheat, oats, and rye these include the testa, pericarp, and various tissues such as glumes and paleae which later become the husk (Fig. 1.18). Initially these tissues make up about 70–80% of grain fresh weight. By maturity their relative contribution has decreased to around 5–10% and they form a tough highly compressed layer around the grain.

The pericarp of these cereals is of some interest since it contains chloroplasts which are active in photosynthesis throughout much of grain development. It is considered that in addition to photosynthesis in the flag leaf, and other tissues of the ear such as glumes and palea, the pericarp may make some contribution to grain dry matter.

As maturation proceeds, dry weight reaches a maximum value and dehydration sets in. The seed shrinks, the outer layers consolidate, and the seed is ready for harvest.

2.1.2 Oilseeds

These include soybean, groundnut, cottonseed and linseed. Although they belong to widely different species (Chapter 1), their developmental morphology has broad similarities. Oddly, while much is known of the developmental characteristics of oilseeds of relatively minor economic importance such as *Sinapis alba* (mustard) and *Crambe abyssinica*, little is known of groundnut or soybean—two of the most popular and economically important.

The oilseeds, when mature, contain high concentrations of unsaturated triglycerides (Section 5.14) which are the source of many of the vegetable oils used in the food industry. The residues left after oil extraction are rich in protein and constitute an extremely valuable by-product (Section 4.3.2).

In oilseed rape, whose developmental anatomy resembles to some extent that of the leguminous seeds, embryo growth within the endosperm is initially slow. Interestingly all cell division in the embryo ceases by about 2 weeks after fertilization, and growth thereafter is by cell enlargement. Thus, the final yield, if it depends on final cell numbers, is probably determined by the first fortnight of seed development. After this, the embryo grows rapidly and finally comprises the major part of the mature seed. The endosperm is used as a nutrient supply for the growing embryo and by maturity all that remains of this tissue is a single layer of aleurone cells. Oil bodies appear in the maturing embryo of *Crambe abyssinica* as early as 8–10 days after petal fall and the increase in triglyceride content of the seed appears to be mainly due to the formation of new oil bodies rather than a large increase in volume of the few original bodies. Oil bodies are fairly small in comparison with other storage organelles, being generally in the range 0.5–1.5 μ m and surrounded by an outer membrane (Fig. 2.3).

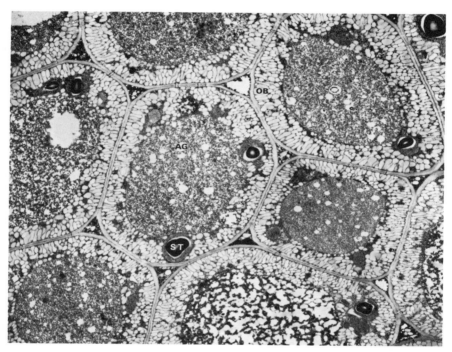

Figure 2.3   Cotyledon parenchyma cell of *Crambe abyssinica* 25 days after petal fall. Aleurone grain (AG); oil bodies (OB); and starch grains (ST). ($\times$ 1 800) (Smith, 1974). (Reproduced by permission of Springer-Verlag)

As in the cereal endosperm, protein deposition is characterized by a steady increase in the number of protein bodies. In the legumes, with the exception of groundnut, these do not contain inclusions, although the seeds probably all contain phytic acid. The protein bodies of linseed, mustard, rape, and cottonseeds contain both globoids and crystalloids and are thus quite distinct from those of soybeans and cereal seed endosperms. Little is known of the development of starch granules during oilseed development, although as much as 12–15% of final dry weight can be starch. Some species, e.g. soybean, do have varieties which lack starch and these are of particular value when formulating diets for diabetics.

40

A particular characteristic of the legumes is that the embryo generally remains green during development. Thus much of the starch may be located in chloroplasts rather than amyloplasts. Certainly the embryo chloroplasts of the developing rape seed produce large starch granules during the early stages of cell expansion. Many of these disappear with the onset of maturity.

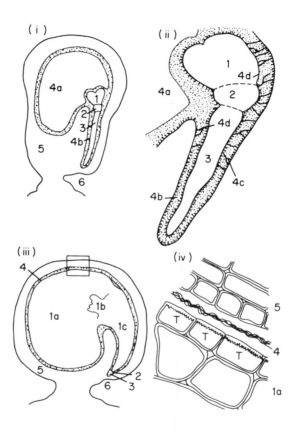

Figure 2.4 Diagrammatic representation of the spatial relationships of endosperm, embryo, and maternal tissues in the developing pea seed (not drawn to scale) (Sutcliffe and Pate, 1977). 1, Embryo; la, cotyledon; lb, plumule; lc, radicle; 2, basal cell; 3, suspensor cell; 4, endosperm; 4a, liquid region of endosperm; 4b, cytoplasmic region of endosperm; 4c, cellulosic strands traversing the cytoplasmic space between embryo and outer wall of embryo sac; 4d, extra embryonic sheath surrounding suspensor and embryo; 5, maternal tissue of seed; 6, micropyle. (i) Early embryo stage at time of maximum development of endosperm. (ii) Enlarged region of endosperm of (i), showing extra embryonic sheath, anchoring strands and shorter wall ingrowths in the endosperm. Details of embryo are omitted. (iii) Stage when embryo first fully fills the endospermic cavity. Note remains of suspensor. (iv) Cellular details of inset of (iii) to show layer of epidermal transfer cells (T) on cotyledon (la), remains of endosperm (4), and nucellus tissue (5). (Reproduced by permission of Academic Press Inc. (London) Ltd)

2.1.3 Peas

Pea embryo development has been well described and follows the pattern broadly outlined above for oilseeds. Fig. 2.4 shows the relationship between endosperm, embryo, and maternal tissues in the developing pea seed. The main phase of accumulation, which commences once the cotyledons have reached their maximum size, is characterized by a rapid deposition of starch and protein. The plastids contain only single starch granules which grow to around 10 μm in diameter. Storage protein is laid down in membrane-bound bodies 1–2 μm in diameter. By maturity all that can be seen is a granular matrix containing masses of protein, starch, and irregular scattered lipid deposits.

2.2 DEPOSITION OF STORAGE MATERIAL IN DEVELOPING SEEDS

The dry matter entering developing barley and wheat grains is derived mainly from photosynthesis in the ear and flag leaf. The relative contributions of the two are variable but it is likely that the ear itself may account for more than 30% of the total grain assimilate. The tissues of the ear which contain chlorophyll and which may contribute to grain assimilate include the awns, glumes, paleae, and pericarp. Controversy surrounds the role of the ear photosynthetic tissues. The awn may make only a minor, or even negative, contribution. Little is known of the other green tissues such as glumes or paleae which surround the developing grains. The green layer of pericarp which surrounds the grain is capable of high rates of photosynthesis, but its position beneath the transparent layer of pericarp (Fig. 1.18) and outer tissues suggest that its role may be confined to the refixation of carbon dioxide derived from endosperm respiratory processes.

About two thirds of the total carbon requirement of pea seeds is met by the photosynthetic organs at the fruiting node, i.e. the pod itself, and the stipules and leaflets of the subtending leaf. The major part of the pod contribution is attributed to the efficient recycling of respired carbon derived from the developing seed. The remaining one third not accounted for presumably comes from elsewhere in the plant (Fig. 2.5).

2.2.1 Nutrients entering the seed

Analysis of the exact composition of nutrients entering the cereal grain has not yet proved possible since these travel in the phloem, a tissue almost impossible to sample in cereals. Sucrose is generally considered to be the source of seed lipid and carbohydrate. The precise source of nitrogen is largely unknown but it is almost certainly not in the form of a complete range of amino acids in the correct proportions. It seems probable that a few amino acids such as glutamate/glutamine and aspartic acid/asparagine

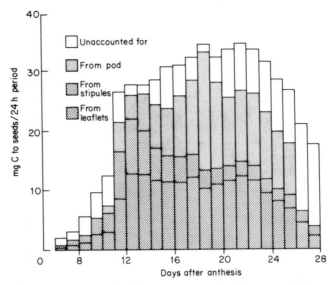

Figure 2.5   A day-by-day record of the provision of carbon to seeds borne at the first fruiting node of field pea by their pod, and by the stipules and leaflets of their subtending leaf. The estimates include photosynthetically fixed carbon, carbon available through mobilization of dry matter, and, in the case of pods, carbon reassimilated from the respiring seeds. These data for transfer are matched against the daily requirement of the seeds for carbon, the latter being calculated by summation of daily losses of carbon by the seeds in respiration and daily gains of carbon by the seed as dry matter. Consequently, the unhatched areas of the histograms (labelled 'unaccounted for') represent carbon required by the seeds but not derived from organs at the fruiting node (Flinn and Pate, 1970). (Reproduced by permission of Oxford University Press)

together with ammonium ion are the main nitrogenous nutrients reaching the developing seeds. These are then subject to much metabolic transformation before deposition of seed proteins. Other substances reaching the grain will include mineral elements and possibly some hormones.

All nutrients are supplied to developing wheat and barley endosperm and embryo via the vascular tissue in the crease. They must first pass through the chalazal region, then the nucellar projection, and finally the aleurone layer before entering the endosperm (Fig. 2.6).

The pea plant is one of the few species from which liquid samples of both xylem and phloem have been analysed. The major solutes of the xylem are nitrogenous and include asparagine, glutamine, and aspartic acid. The xylem also contains growth substances and mineral ions. The major constituent of the phloem stream is sucrose, the remainder including small amounts of other sugars and a range of amino acids. Again, as with the developing cereal grain, considerable metabolic changes must take place in the developing pea seed before the deposition of reserve materials can take place.

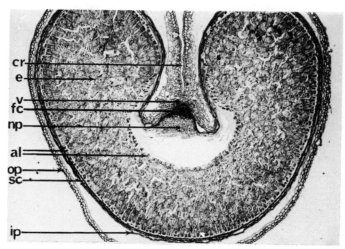

Figure 2.6 Transverse section through a developing wheat grain 14 days after anthesis at mid-point between base and apex. cr, Crease, e, starchy endosperm; v, vascular bundles; fc, funiculus–chalazal region; np, nucellar projection; al, aleurone layer; op, outer pericarp; sc, seed coat; ip, inner pericarp (Sakri and Shannon, 1975). (Reproduced by permission of the American Society of Plant Physiology)

2.3 CARBOHYDRATE DEPOSITION IN CEREALS

2.3.1 Sucrose utilization and starch biosynthesis

In rice, barley, and sweet corn endosperms it is considered that the sucrose entering the grain is mainly metabolized by UDP-dependent sucrose synthase (Fig. 2.7). In the early stages of maize endosperm and embryo development, however, some sucrose may be cleaved by invertase. Thus, the products of sucrose metabolism are mainly UDP-glucose with some ADP-glucose, free glucose, and fructose.

The preferred nucleotide sugar donor for starch synthase is ADP-glucose. Thus it may be that UDP-glucose, which is formed first, is converted to glucose-1-phosphate via UDP-glucose pyrophosphorylase and thence to ADP-glucose by a reversal of ADP-glucose pyrophosphorylase. This mechanism is incorporated into Fig. 2.7, which summarizes the possible pathways of starch biosynthesis.

In the past, phosphorylase was thought of as the enzyme responsible for amylose synthesis, but starch synthase is now generally considered to be the enzyme responsible for the bulk of starch synthesis although phosphorylase may have a role in the synthesis of primers for starch synthase, particularly in the early stages of grain development.

The presence of starch granules of various sizes and compositions (Section 2.1.1) suggests that there may be several amyloplast-bound synthases.

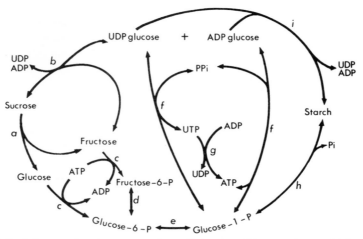

Figure 2.7 Scheme showing a number of the possible reactions which may be involved in the conversion of sucrose to starch in developing endosperms. a, invertase; b, UDP(ADP)-dependent sucrose synthase; c, hexokinase; d, glucose-6-phosphate ketoisomerase; e, phosphoglucomutase; f, UDP(ADP)-glucose pyrophosphorylase; g, nucleoside diphosphokinase; h, phosphorylase; i, ADP(UDP)-glucose starch synthase (Perez *et al.*, 1975). (Reproduced by permission of the American Society of Plant Physiologists)

However, work with mutant maize endosperms has made it clear that the nature of the genetic, biochemical, and possibly environmental factors that control the size, shape, and composition of starch granules is largely unknown. In particular the biosynthesis of amylopectin, which comprises 75% of most cereal starches, has been subjected to little investigation—largely because of difficulties in assaying branching enzyme in the presence of interfering amylases (Section 2.3.2).

The starch content of cereal grains increases steadily throughout development, accounting for around 65–70% of grain dry weight at maturity. In general it seems that the amount of amylose relative to the total amount of starch present also increases throughout grain development. That is, in the early stages of grain development, amylose accounts for as little as 10% of the total starch present. The proportion then rises steadily to reach a final value of around 25–30%.

In a number of cereals including barley, wheat, and maize it seems that the larger starch granules have a significantly higher amylose content than the smaller granules. Thus the changes in starch composition as development proceeds may be due to changes in the relative numbers of granules of different sizes as well as to changes in the individual amylose contents of all granules.

2.3.2 Limitations to starch biosynthesis

A considerable amount of carbohydrate may be lost by grain respiration.

This results from catabolic processes active in carbohydrate degradation during endosperm development. Obviously, intermediates and energy are required for starch and protein synthesis and presumably these are supplied by respiration. Both glycolytic and mitochondrial enzyme activities are present in immature grains together with a number of starch-degrading enzymes such as phosphorylase, α- and β-amylases, and debranching enzyme.

However, immature starch granules are only slightly degraded by cereal amylases, indicating that the substrate for respiration may be sucrose rather than the insoluble starch reserves.

Starch synthesis in wheat and probably barley and maize may be limited not by sucrose supply but by regulatory mechanisms within the grains themselves; for example, low ADP-glucose pyrophosphorylase activity, or simply the size and number of endosperm cells. The presence of grain catabolic processes, such as respiration, might be expected to lower yields. So far, however, no relationship has yet been demonstrated between amylase activity and final grain yields. The ability of the pericarp to refix carbon dioxide, which varies substantially between cultivars, may be an important additional factor in controlling losses and maximizing starch deposition.

2.3.3 Non-starch polysaccharides

The principal polysaccharides of cereal grains, other than starch and fructan, are cellulose and hemicelluloses. Cellulose is confined mainly to the outer layers of the grain, i.e. the husk, whereas hemicelluloses can be found throughout.

The endosperm cell wall is unique among those of higher plants since it contains no pectin and very little cellulose. Much is composed of β-glucan, which is a linear molecule with 30% β-(1,3)- and 70% β-(1,4)-linkages and associated in the cell wall with firmly linked peptide sequences. Interest in cell wall synthesis stems from the fact that excessive release but incomplete hydrolysis of β-glucan during malting can result in slow filtration during wort run-off and lead to haze formation in high-gravity beers. Unfortunately a successful screening procedure for selecting barleys for low rates of β-glucan synthesis has not yet been developed.

2.4 CARBOHYDRATE DEPOSITION IN SEEDS OTHER THAN CEREALS

Little is known of the factors controlling carbohydrate deposition in seeds other than the cereals—probably because the amounts present are not sufficient to justify the setting up of a research programme. Of such seeds the pea has been subject to most investigation since the principal storage material of the cotyledons is carbohydrate. In the garden pea (*P. sativum*) this is mainly starch, but in the field pea (*P. arvense*) hemicelluloses predominate.

Obviously the main interest in the carbohydrate metabolism of developing garden pea seeds stems from the fact that palatability is related to the sucrose content. The same is true of sweet corn kernels. The sucrose content of the pea seed decreases once starch synthesis becomes appreciable and so harvesting, which in any case is a compromise between size, succulence, and tenderness, depends on a balance between the rates of sucrose supply and starch synthesis. The biochemistry of starch synthesis in developing peas is probably very similar to that in cereals. However, in the field pea other enzymic mechanisms must exist to divert the sucrose into the pathways of hemicellulose metabolism. The factors controlling this have not been investigated.

2.5 PROTEIN DEPOSITION IN SEEDS

The protein reserves of cereals are mainly found in the endosperm with a small amount, around 2–5%, in the embryo. In contrast, there is usually little or no endosperm in mature legume seeds and the protein reserves are almost entirely confined to the cotyledons. The final protein content depends on a number of factors including variety, fertilizer inputs, and environmental conditions such as light, temperature, and water availability during plant growth and seed maturation. For example, in cereals, early application of nitrogenous fertilizer can increase overall yield but not the amount of protein per grain. This is exploited in the production of malting barleys where high yield coupled with a low nitrogen content is required (Section 2.5.5). The relationship between nitrogenous fertilizer application and crop yield is generally considered to be hyperbolic and late applications of nitrogen can increase cereal grain protein content considerably. Since this is due, however, to increased synthesis of the prolamins and glutelins (Section 2.5.2), the protein quality is much reduced.

A major problem confronting crop physiologists is to combine high grain yield with increased, and nutritionally better, seed protein. Unfortunately, it seems to be generally the case that seed protein content is inversely correlated with yield (Fig. 2.8). This, together with the negative correlation between seed protein content and protein quality, poses a formidable obstacle to progress.

2.5.1 Amino acid synthesis and supply

Little is known of the physiological factors regulating the supply of nitrogen to the seed, or of the pathways involved in its incorporation into protein, or indeed into any of the wide variety of nitrogenous compounds found in seeds, which include nucleic acids, vitamins, and coenzymes.

While undoubtedly some of the amino acids supplied to developing seeds will be incorporated directly into protein, it is probable that considerable synthesis of amino acids takes place within the seeds themselves in view of the

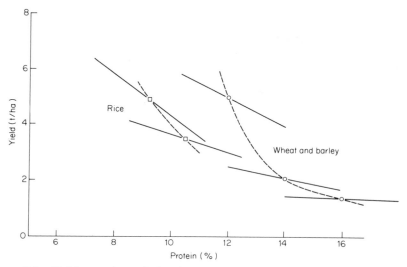

Figure 2.8 Yield–protein relationships in cereals. Wheat and barley agree remarkably closely and can be shown on one set of curves; rice is different (lower protein content); other cereals are not well enough known to be included. The solid lines are genetic regressions, the broken lines generalized environmental relations; the higher is the yield the steeper are the slopes of the genetic regressions. (Reproduced by permission of Professor N. W. Simmonds)

high levels of amides and general imbalance of amino acids in the phloem. Certainly, enzymes capable of catalysing amino acid synthesis are present in developing seeds. For example, glutamate synthase, which can utilize glutamine and α-ketoglutarate in the synthesis of the amino acid glutamate, is present in developing maize endosperm and pea cotyledons (Fig. 2.9).

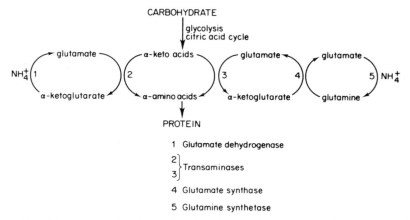

Figure 2.9 Enzyme-catalysed reactions involved in the synthesis of amino acids from carbohydrate and ammonium ion

48

Glutamine is derived from the phloem stream and α-ketoglutarate presumably is a product of carbohydrate respiration. Glutamate synthase activity has been positively correlated with nitrogen accumulation in developing maize (Fig. 2.10). The glutamate formed could subsequently be used in the synthesis of other amino acids by means of transaminase enzymes known to be present in immature cereal endosperms, and doubtless also in most developing seeds. Any ammonium ions reaching the seed may be metabolized by either glutamate dehydrogenase or glutamine synthetase to give the amino acids glutamate and glutamine, respectively (Fig. 2.9).

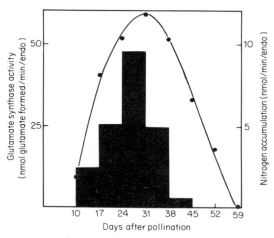

Figure 2.10 Glutamate synthase activity (●) and rate of nitrogen accumulation (histogram) in maize endosperm during development (Sodek and da Silva, 1977). (Reproduced by permission of the American Society of Plant Physiologists)

It is generally assumed that the pathways of amino acid synthesis and metabolism in developing seeds are similar to those in other living tissues. Certainly the transaminases which utilize glutamate for the synthesis of aspartate and alanine are present in developing barley grain. In turn, aspartate is itself a precursor of the essential amino acids threonine, methionine, lysine, and isoleucine. The aromatic amino acids tryptophan, tyrosine, and phenylalanine—also essential amino acids—are probably derived from the carbon substrates erythrose-4-phosphate and phosphoenol pyruvate, which are intermediates of carbohydrate degradation. The nitrogen comes either from glutamine or from other amino acids by transamination.

2.5.2 The seed proteins

Protein is the major nitrogenous reserve material of seeds and, since it is likely that during development much of the protein synthesized has originally a physiological role, there are probably many hundreds of different proteins in the seed. With such a variety of complex macromolecules it is difficult and probably not even desirable to study the deposition of individual proteins, but preferable to study the control of deposition of groups of proteins with similar properties. Such groups were originally classified by Osborne (1924) on the basis of their solubility and this method is still in use today. Since seeds differ greatly in their nutritional and biochemical properties, it has enabled a direct comparison to be made between the various protein fractions.

There are four solubility groups. Each is highly heterogeneous and varies from species to species: albumins are soluble in water in neutral or slightly acidic conditions; globulins are soluble in salt solutions but insoluble in water; glutelins are soluble in strong acid or alkali, but insoluble in water, salt solutions or ethanol; and prolamins are soluble in 90% ethanol but insoluble in water.

Albumins are presumably soluble enzymic proteins and would include such enzymes as sucrose synthase, starch synthase, and transaminases. Globulins are also highly heterogeneous and in legumes can be separated into two major fractions, legumin and vicilin. The globulins are minor constituents of the cereal grains. However, in most leguminous seeds they account for well over 60% of total protein. The glutelins are complex proteins of high molecular weight and include hordenin of barley endosperm and glutenin of maize. Rice protein in particular has a high glutelin content. The prolamins are uncommon in seeds other than those of the Gramineae and include gliadin of wheat, zein of maize, and hordein of barley. The relative amounts of the various protein fractions in a number of different seeds are shown in Table 2.1.

Table 2.1. Protein fractions of various seeds (Beevers, 1976; Norton, 1978). (Reproduced by permission of Edward Arnold (Publishers) Ltd. and Butterworths)

	Albumin	Globulin	Glutelin	Prolamin
Wheat	5	10	40	45
Maize (normal)	14	—	31	48
(opaque-2)	25	—	39	24
Rice	5	10	80	5
Pea	21	66	12	—
Mungbean	4	67	29	—
Soybean	10	90	—	—
Groundnut	15	70	10	—

The variation in solubility of the different protein fractions is due to differences in their amino acid composition (Table 2.2). Protein quality, from a nutritional point of view, is optimal in the albumins. In particular, they have the highest content of the usually limiting amino acids lysine, tryptophan and methionine (Section 4.1.1). The globulins are also of good quality, being enriched in the essential amino acids arginine and aspartate as well as lysine. Differences, however, are observed when the globulins from different species are compared (Table 2.2).

Table 2.2. Amino acid composition of some reserve proteins from barley, soybean and maize (g amino acid per 100 g protein) (Beevers, 1976). (Reproduced by permission of Edward Arnold (Publishers Ltd)

Amino acid	Albumin	Globulin		Glutelin		Prolamin	
	Barley	Barley	Soybean	Barley hordenin	Maize glutelin	Barley hordein	Maize zein
Arg	6.5	11.0	7.8	6.0	5.0	3.0	2.1
His	2.5	1.8	2.4	2.5	4.4	1.3	1.4
Lys	6.7	5.3	6.0	4.0	2.1	0.7	0.1
Try	1.5	0.8	3.8	1.3	5.6	0.8	5.6
Phe	5.1	2.8	5.5	3.6	5.1	3.0	7.6
Cys	2.1	3.6	1.2	1.2	1.4	2.1	1.0
Met	2.4	1.5	1.2	1.9	2.4	1.3	1.7
Ser	4.9	4.7	5.7	5.0	5.5.	3.8	6.2
Thr	4.6	3.3	3.6	4.2	4.2	2.6	3.2
Leu	8.6	6.8	7.9	8.7	12.9	6.9	22.4
Isoleu	6.2	3.3	4.8	5.2	3.3	5.4	4.2
Val	7.8	5.5	4.7	6.6	4.3	4.7	4.5
Glu	12.9	11.9	22.4	19.8	22.0	39.6	27.4
Asp	12.2	8.5	12.6	7.1	5.4	1.8	5.9
Gly	5.7	9.2	4.1	4.5	4.5	1.5	1.6
Ala	7.3	0.7	3.9	6.7	7.2	2.2	10.9
Pro	5.5	3.6	5.4	8.7	13.9	20.1	10.7

The glutelins, which are particularly characteristic of rice, are intermediate in quality between the albumins and prolamins. The poorest quality protein is that of the prolamins. These are particularly poor in lysine and other essential amino acids such as arginine and histidine. The generally small amounts of essential amino acids are compensated for by increased glutamine and proline contents.

2.5.3 Developmental aspects of protein deposition

The albumins are present in small amounts in most mature seeds (Table 2.1) but normally account for a high percentage of total protein at the very earliest stages of development. For example, in developing pea cotyledons protein accumulates rapidly between 12 and 25 days after fertilization (Fig. 2.11) with the albumins as the first proteins synthesized followed by the globulins, initially the soluble vicilin, and then insoluble legumin. Similarly, during development in barley endosperm, more than 50% of the albumins are already present by 8 days after fertilization. Very small amounts of the globulins, hordeins (prolamin), and glutelins are formed during the first 13 days but by 20 days their rates of synthesis reach a maximum value.

In common with many of the biosynthetic systems of developing cereal grains, the incorporation of amino acids into maize endosperm protein is initially slow, but increases to a maximum roughly half way through the developmental period, and then declines. A number of recent studies have

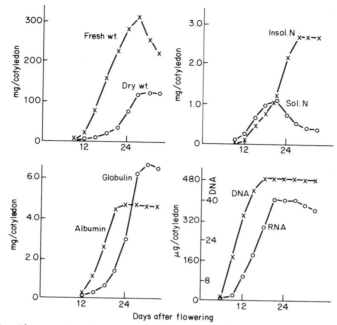

Days after flowering

Figure 2.11   Changes in nitrogenous components of pea (*Pisum sativum*) cotyledons during development (Beevers, 1976). (Reproduced by permission of Edward Arnold (Publishers) Ltd)

been concerned with the synthesis of specific storage proteins such as zein. It has been suggested that unique classes of large membrane-bound polysomes bound to the outer surface of zein-containing protein bodies are the principal sites of zein synthesis. Thus it appears that zein is synthesized conveniently at its site of deposition within the maize endosperm.

In oil-bearing seeds such as cotton and groundnuts the reserve proteins may be synthesized on the polysomes of the rough endoplasmic reticulum. It is believed by some workers that the proteins are passed through the membranes to the lumen of the endoplasmic reticulum and then transported to the Golgi apparatus where they are concentrated in membrane-bound droplets. These are then postulated to migrate through the cytoplasm to the vacuolar space and the protein is transferred to the vacuole by means of membrane fusion. It is interesting that those seeds which concentrate most protein should apparently have such a complex system for protein deposition whereas the cereals, with much smaller amounts of protein in the grain should have a system which seems to be comparatively simple.

During the early stages of maturation in soybean a low molecular weight protein fraction predominates and subsequently decreases. This is presumably the labile pool of enzymic proteins or albumins. The vicilins and legumins are synthesized later in maturity but accumulate steadily to become the major proteins of the mature seed.

Nucleic acid metabolism in developing seeds has also been subjected to some investigation. In developing pea seeds DNA accumulation commences early and continues even after cotyledon cell division has terminated. Levels of DNA and RNA remain roughly constant once dehydration has commenced. In developing maize endosperm, on the other hand, maximal DNA content is thought to coincide with the termination of nuclear and cell division. It may then be that the higher protein content of legumes is due, at least in part, to the relatively longer period of DNA synthesis.

Some success has been obtained in the *in vitro* synthesis of seed storage proteins using cell free systems. For example, the messenger RNAs coding for zein synthesis have been isolated from immature maize endosperm and shown to be capable of directing zein polypeptide synthesis using polysomes derived from wheat germ.

2.5.4 Mutant or 'improved' genotypes

Numerous attempts have been made, particularly with cereals, to improve seed protein quantity and quality by genetic manipulation. A range of 'improved' genotypes has been discovered using either exhaustive screening procedures or chemically- or radioactively-induced mutations. On the whole, however, such plants have poor agronomic characteristics, in particular low yield, but frequently also poor grain conformation and lowered disease resistance. In spite of this, much work is being carried out on their unique biochemical and morphological properties. For example, in maize, the mutants opaque-2 (02) and floury-2 have an improved amino acid composition which is due to reduced zein synthesis and increased glutelin synthesis during maturation (Nelson). This results in an increase in the grain content of lysine, tryptophan, aspartate, and glycine and decreases in the content of leucine, alanine, and glutamate. The improvement in seed protein quality is associated with a reduction in size of the protein bodies to around 0.1 μm. Endosperm protein synthesis in the mutant opaque-6 (06) is very similar to that in 02, that is a reduction in zein synthesis with an increase in free amino acids, albumins and globulins. Thus the amino acid pattern of the triple 06/06/06 endosperms is similar to those of 02/02/02 endosperms. Yet 06/06 plants die as seedlings.

The lethality of a mutant where apparently only grain storage protein is affected is difficult to explain. Possibly storage protein has functions other than as a source of readily available nitrogen to the growing embryo. Perhaps the mutation may depress synthesis of an enzyme or other protein required but not synthesized *de novo* during germination and early development.

A range of barleys with a higher than normal lysine content have also been identified in different parts of the world. Their major property is a decreased hordein synthesis during grain development. These include the variety Hiproly, which was found by screening barleys from the world collection for high lysine content, and the Risø mutant 1508 derived from the variety Bomi by induced chemical mutation using ethyleneimine. In a study of Hiproly grain development, lysine levels were initially similar to the control (low lysine barley) but increased faster as development proceeded—presumably owing to the decreased hordein synthesis and increased synthesis of fractions within the albumin/globulin fraction of the protein. Unfortunately, Hiproly has particularly poor agronomic characteristics which limit the possibility of its commercial exploitation as a parent variety. The 1508 mutation not only affects protein synthesis but also has devastating effects on both the general biochemistry and morphology of endosperm development. In addition to reduced hordein synthesis there is a reduction of lysine-poor components of the glutelin fraction and a compensating increase in the amount of lysine-rich glutelins as well as free lysine (Fig. 2.12).

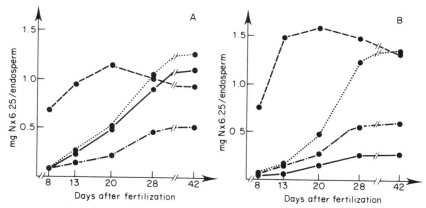

Figure 2.12   Changes in endosperm protein fractions during kernel development of Bomi barley (A) and mutant No. 1508 (B) Albumins (including free aminoacids) – – – – –, globulins –·––·–, hordeins ——————, and glutelins .... (Brandt, 1976). (Reproduced by permission of the American Association of Cereal Chemists)

While the lysine content of the mutant is about 40% higher than its parent Bomi and the protein content of the grain about 7% higher (Table 2.3), the overall yield of grain is about 10–15% lower than that of commercial varieties. In addition, the seeds have a characteristically shrunken endosperm and are smaller in size and weight than normal varieties. This is largely due to a reduction in starch synthesis. Thus the mutant has the further disadvantage of producing a lower metabolizable energy (Section 4.1.2) when fed to animals. It must be emphasized that a high lysine seed is one which has a higher lysine content than normal varieties with the same protein content. If a high lysine type has also a higher protein content, its protein quality (Section 4.1.1) may well be lower than that of the normal variety since it may actually have less lysine in the protein. This is seen in Table 2.3, where the protein and lysine contents of the parent Bomi and several of its mutants are shown. Since the protein content of the mutants is between 7 and 24% higher than the parent the increase in lysine would be somewhat higher if they were compared with normal barleys having the same protein content. Since there is little evidence that the expression of the shrunken character can be modified by breeding, the prospects for Risø 1508 and probably many others seem limited. With legumes and other seeds, however, the technique of crop improvement using artificially induced mutations has been little used. Most progress has been achieved by more conventional techniques (Section 2.5.5).

Table 2.3. Grain yield and composition of the barley variety Bomi compared with its high lysine mutants (Doll, Køie and Eggum, 1974). (Reproduced by permission of Pergamon Press)

Type	Grain yield (g/plot)	Seed size (mg)	Protein content of grain (%)	Lysine per 16 g N (g)
Bomi	492	45	10.4	3.73
Mutant 440	316	30	12.8	4.32
Mutant 527	369	39	11.5	4.18
Mutant 1508	434	41	11.1	5.30
Standard error	19	1	0.3	0.05

2.5.5 Improvement of protein yield and quality

In addition to the exploitation of mutants, much emphasis is placed on the improvement of seed quality by conventional plant breeding techniques. This is mainly directed at increases in seed protein content as well as increases in the amount of certain essential amino acids in that protein, notably lysine and the sulphur-containing amino acids methionine and cystine. Of course, the simplest method of increasing protein production is to improve gross yield, where the seed protein content remains the same or even slightly less than normal. However, if the amounts of protein and essential amino acids per seed are increased then animals would have to eat less on a weight basis to satisfy their protein requirements. Additionally, by consuming less carbohydrate and more protein, especially in cereals, they would increase their muscle to fat ratio, a factor of particular importance when preparing animals for slaughter.

As described previously (Section 2.5.4), it seems that with cereals, pulses, and oilseeds simultaneous increases in seed protein concentration and yield may be incompatible, particularly in modern cultivars under high-yielding conditions where photosynthetic capacity and sink size may be closely balanced. More photosynthate may be required to synthesize seed protein than seed carbohydrate, as to synthesize 1 g of protein requires in theory (based on guesswork of the probable seed biosynthetic pathways) around 1.7–2.5 g of glucose, whereas to synthesize 1 g of seed carbohydrate requires only 1.2 g of glucose. Thus, for example, to produce seeds with a 1% higher protein content but with no increase in overall weight would require approximately a 1% increase in the supply of photosynthate. This assumes the most unfavourable case where photosynthesis and sink size are just balanced, and where the increased requirement is not associated with an increase in the supply of photosynthate to the developing grain. However, the typical yield reduction in improved protein cereal cultivars exceeds the predicted reduction in most cases. It may then be that the genes responsible for high protein content and quality are associated with reduced deposition of carbohydrate in the endosperm. These effects are much less for peas and for low yielding cultivars of rice and barley where, in addition to other factors, photosynthesis may be well in excess of requirements.

Obviously nitrogen, too, may be a limiting factor in increased protein yields. From Table 2.4 it is clear that each 1% increase in grain protein would give rise to an increase of 6–11% in the nitrogen requirement. Alternatively, the possibility of introducing a nitrogen fixing capacity into cereal grains exists. Even then, however, this has an energetic cost and may again compete with grain filling for supplies of photosynthate.

Table 2.4. Standard chemical compositions and nitrogen requirements (mg N per g photosynthate) for cereal grains. The final column gives the percentage increase in nitrogen requirement* for a 1% increase in protein. (Bhatia and Rabson, 1976). (Reproduced with permission from *Science,* 1976, **194,** 1418–1421. Copyright 1976, American Association for the Advancement of Science)

| Crop | Assumed standard composition (% dry matter) | | | | Nitrogen requirement (mg/g) | | Increase in nitrogen requirement (%) |
	Protein	Carbohydrate	Lipid	Ash	With standard protein	With 1% increase in protein	
Wheat	14	82	2	2	16.0	17.0	6
Rice	8	88	2	2	9.7	10.8	11
Maize	10	84	5	1	11.3	12.3	9
Barley	9	80	1	4	11.5	12.3	10
Sorghum	12	82	4	2	13.6	14.6	7
Oats	13	77	5	5	14.8	15.8	7
Rye	14	82	2	2	16.0	17.0	6

*The additional nitrogen required is calculated by assuming that protein is 16% nitrogen by weight.

Additional problems are encountered in such seeds as the oilseeds where both the protein and the oil are important products. In seeds of the same weight there is obviously a negative correlation between oil and protein content, and so increases in protein must be made at the expense of the oil content. Thus, the relative demand for these two products dictates whether or not an increase in seed protein content is desirable. Clearly, new and more specialized oilseed varieties are required which can be grown in response to market requirements.

There is less interest in improving the protein content of those seeds which already contain more than 20% protein. These include the legumes and some oilseeds. It is generally considered that the improvement of overall yield, protein quality, and agronomic characteristics in these plants should be the first consideration in any development programme. There is a wide range of legume seed protein content: from 31 to 52% for soybeans and from 15 to 35% for peas. However, these values are influenced significantly by the environment. It is a major problem in breeding for higher protein content to distinguish genetic from environmental effects.

Improvements in protein quality most likely result from changes in the relative amounts of the various protein fractions present. However, the albumin content, which is of highest nutritional quality, of cereals is low and any improvement in overall quality can result only from fairly drastic increases in the proportion of this fraction. Reduction in prolamin deposition, as we have seen in the previous section, gives an improvement in quality, but at the expense of overall yield and even total protein.

In breeding food legumes for improved protein quality most effort has been directed at screening for amino acid variability and determining if amino acid content is genetically controlled. One of the barriers to progress has been the lack of a satisfactory method for measuring methionine in large numbers of samples. Both methionine and cystine are the limiting amino acids of legumes. However, the genetic relationship between yield, protein content, and methionine content has still to be evaluated. For example, with mung beans, the methionine content ranges from 0.55 to 1.78% of the protein, but those strains highest in methionine have a reduced protein content.

It might well be that the most efficient course of action is simply to breed for higher yields and supplement the products with the nutrients in which they are deficient (Section 4.1).

The improvement in nutritional quality of seed protein is not the sole objective of breeding programmes. For example, barley breeders in Scotland are attempting to produce new barleys with high malting potential in order to reduce the almost complete dependence of the Scottish grain whisky industry on English and Canadian varieties. The programme involves both conventional breeding techniques and chemically induced mutagenesis. The objective is to produce high yielding varieties which give high levels of starch-degrading enzymes on malting (Section 5.2.4). Furthermore, since starch-degrading activity is observed only in the lysine-rich fractions, i.e. the albumins and globulins, the quality of grain protein should also be improved.

In wheat the baking quality required for leavened bread depends on the physical properties of the grain proteins (Section 5.6.1). Since this in turn is correlated to some extent with protein content, breeding programmes are concentrating on producing a wheat which will yield at least 12% protein under variable environmental conditions, combined with other favourable processing and agronomic characteristics.

2.5.6 Improvements in quality by altered seed morphology

Another possibility for improving seed quality is to select for seeds with increased proportions of those tissues which have relatively the highest nutritional or other properties. For example, it might be worth considering selecting cereal grains for a higher embryo content since on a dry weight basis these have a higher protein content and quality than their associated endosperm as well as a higher content of lipids and some vitamins (Table 2.5). These increases are, of course, at the expense of carbohydrate. Such grains would not be suitable for malting, in the case of barley (Section 5.2.6), or breadmaking (Section 5.6.1), in the case of wheat. However, the embryo comprises only around 5% of the dry weight in cereals, so that drastic changes in morphology would be necessary in order to achieve a significant increase in nutritional value.

Table 2.5. Relative composition of embryos and kernels in maize. (US–Canadian Tables of Feed Composition, 1969). (Reproduced by permission of the National Academy of Sciences). Figures expressed per kg dry matter.

Component	Kernel*	Embryo†
Ether extract (g)	44	82
Crude protein (g)	102	217
Calcium (mg)	300	500
Phosphorus (mg)	3 100	5 500
Iron (mg)	30	400
Riboflavin (mg)	1.3	3.3
Thiamine (mg)	4.6	21.3
Niacin (mg)	26.6	43.8

*Zea mays var indentata (dent yellow).
†Maize germ meal, mechanically extracted (wet-milled).

Again, in cereals, an increased thickness of the aleurone layer which is rich in basic amino acids and lipid would undoubtedly improve the nutritive value. However, the increased amounts of phytic acid (Section 2.7.1) might outweigh any advantages conferred. A further possibility is to reduce the proportion of the outer seed layers which are largely indigestible to man and other simple stomached animals due to their high cellulose and hemicellulose content. The advantage of such changes is open to some doubt since recent work suggests that dietary fibre may have a beneficial effect on digestion in man. It is also considered that the outer seed layers give protection against damage and infection during seed growth and storage. Naked varieties of barley are known but have not made much impact on commercial practice.

2.6. LIPID DEPOSITION

The widest definition of the term 'lipid' includes all those naturally occurring materials which are soluble in organic solvents. Since this involves a wide range of often unrelated substances from steroids to fat-soluble vitamins, the present discussion will be confined to those present in greatest amount, i.e. to those lipids containing fatty acids and generally called acyl lipids.

Plant lipids are more highly unsaturated than those derived from animals. This is due to the relatively high proportion of unsaturated fatty acids such as oleic, linoleic, and linolenic acids. Animal fats, particularly of ruminant animals such as the cow or sheep, contain a much higher proportion of saturated fat in which stearic and palmitic acids (both saturated fatty acids) predominate.

Plant seed oils are of enormous economic importance in the food industry, and supply about 60% of the world's consumption of oils and fats, with the remainder coming from animal fats and marine oils. The high degree of unsaturation of seed lipids means that most are liquid or semi-solid at temperatures above 20 °C, thus facilitating extraction before subsequent processing.

Those seeds which contain economically important reserves of lipid (oil) include not only the group traditionally referred to as the oilseeds, but also certain high oil lines of such cereals as maize and rice. Compared with other grain legumes such as groundnut (peanut) and soybean, the pea does not synthesize or store significant amounts of lipid in its seeds. Final concentrations of lipid in pea seeds are around 1% of seed dry matter.

2.6.1 Oilseeds

The oil content of these seeds varies considerably both within and between species. Soybeans have probably the lowest and groundnut the highest of those which are commercially important (Table 5.3). The major fatty acids in the lipid of the most economically important seeds are fairly similar. For example, cottonseed, groundnuts, and sunflower contain both oleic and linoleic acids as major fatty acids with linolenic acid also being present in soybeans and in 'zero-erucic acid' rapeseed. The fatty acid composition of some common seed-oils is shown in Table 5.3.

During the initial stages of seed development in a number of oilseeds, lipid deposition is very slow and is accompanied by a substantial but temporary accumulation of carbohydrate. Since seed lipid is generally considered to be a secondary product, derived from sucrose transported from the plant's photosynthetic organs, this is not altogether surprising.

The biochemistry of lipid synthesis in seeds has not been studied in depth and it is assumed to be similar to that prevailing in animals, the major difference being that active desaturase systems resulting in linoleic, linolenic, and oleic acid synthesis are present. Triglyceride synthesis then results from the combination of fatty acyl CoA derivatives with L-α-glycerophosphate derived from the pathways of seed respiration.

It seems that, in most developing seeds, including the cereals, the lipids initially synthesized are principally phospho- and glycolipids, with triglyceride predominating only at later stages of development (Fig. 2.13). These results correlate well with the observed proliferation of membranes during early development and the later appearance of oil storage bodies which presumably are the site of triglyceride deposition (Section 2.1.2). In soybean, for the first 18 days after flowering, the total lipid remains steady at around 1% of total bean weight. This compares with 0.4% in a high-oil variety of maize (Illinois) and 0.06% in a normal variety of barley at a similar stage of development. Thus, even at early stages of development before the appearance of oil bodies, soybean and probably other oilseeds have levels of lipid considerably greater than those found in non-oil-storing seeds. However, by 18 days after flowering, when the oil bodies first appear, 46% of total lipid is triglyceride and by maturity this has increased to 88% (Fig. 2.13).

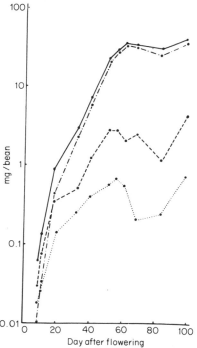

Figure 2.13 Changes in major lipid groups, expressed on a logarithmic scale as weight per bean, in developing soybeans (Privett *et al.*, 1973). ————, Total oil; —·—·—·—·, neutral lipids; – – – – –, phospholipids; ...., glycolipids. (Reproduced by permission of the American Oil Chemists' Society)

The fatty acid composition of soybean lipid also changes markedly during development. For example, the saturated fatty acids palmitic and stearic acid form a relatively high percentage of early formed lipids (38 and 22%, respectively), falling off rapidly thereafter. Oleic and linoleic acids increase rapidly throughout while linolenic acid rises rapidly at first, but subsequently decreases. Such observations may be consistent with the synthesis of unsaturated fatty acids from saturated precursors.

A similar overall pattern of lipid accumulation is found for *Crambe*, a relatively new oilseed crop, first tested in the early 1930s in the USSR, and

Brassica napus, a swede rape well known in Europe, Japan, Chile, and Canada. Major differences are, however, seen in the pattern of accumulation of the fatty acids. By 2 weeks after flowering in both *Crambe* and rapeseed, fatty acid synthesis is limited almost entirely to erucic acid. This synthesis can be suppressed, however, and no erucic acid is synthesized during development in 'zero-erucic acid' varieties of rape. Since erucic acid is toxic these were bred in response to increased demands for new sources of plant seed-oils.

The site or sites of fatty acid synthesis are not precisely identified. In *Crambe*, the oil bodies themselves have been identified as the site of erucic acid synthesis, whereas in safflower seeds microsomal preparations contain the enzyme system for linoleic acid synthesis. Possibly there is co-operation between the two in the overall process of triglyceride deposition.

2.6.2 Cereal grains

The oil content of cereal grains at maturity, much of which is in the embryo, varies both within and between species. For example, oats might contain as much as 8% of dry matter as lipid but wheat as little as 2%. Some varieties of maize, e.g. Illinois high oil (Table 2.6), may reach lipid contents as high as 14% of dry matter—a figure which is probably a maximum value for cereals. Normal values for cereals are around 2.6–5%. The major class of lipid synthesized is triglyceride, although, as with all the other seeds, polar lipids are those first synthesized. The relative amounts of linoleic, linolenic, and palmitic acids of the triglyceride fall throughout maturation while that of oleic increases. Even so, at maturity the major fatty acid present is linoleic acid. In developing wheat, as with maize, the main fatty acids are palmitic,

Table 2.6. Accumulation of lipids in developing high- and low-oil lines of maize (Weber, 1969)

Strain	DAP*	Dry wt. 100 kernels (g)	% oil dry wt.	Triglycerides†	Polar lipids†
Illinois high oil	10	1.1	3.0	10.1	70.8
	15	2.5	5.6	41.1	45.1
	30	11.5	10.9	78.4	9.6
	45	18.0	13.7	84.0	6.0
	60	19.0	13.8	88.1	4.5
	75	23.4	13.4	92.0	3.9
	90	23.8	13.8	92.4	3.9
Low-oil line	10	0.7	3.4	17.3	72.4
	15	1.7	3.6	27.8	63.1
	30	11.0	2.9	56.3	33.1
	45	17.7	3.9	74.4	13.9
	60	20.3	2.4	75.8	13.0
	75	16.5	2.6	74.9	12.0

*Days after pollination
†Percentage of total lipid.

oleic, linoleic, and linolenic acids. Again, linoleic acid is the major fatty acid present throughout development and, again as with maize, triglyceride forms over 70% of the total lipid fraction by maturity.

2.7 MINERAL DEPOSITION

Obviously the multitude of metabolic processes in developing seeds requires the presence of a number of mineral ions for activity. Interestingly, very little is known of the mechanisms regulating ion uptake into developing seeds and whether or not these limit, in any way, the final levels of storage materials.

It seems clear that in developing barley and pea seeds translocation of minerals may occur between the different organs associated with the developing seed. For example, in barley at 7 days post-anthesis 90% of grain zinc is associated with the testa–pericarp, but by 55 days this falls to around 5%. These losses coincide with striking gains in endosperm and embryo zinc. Similar losses and gains can be observed for the normally mobile elements potassium and magnesium. Comparison of similar data for developing pea pods and their seeds show that the period when the pod is losing its mineral reserves most rapidly coincides exactly with the time of fastest gain of mineral ions by the seed. The proportion of seed requirements during development that might be supplied by the pod is shown in Table 2.7.

Table 2.7. Senescence losses of minerals from the pod of *Pisum sativum* (cv. Greenfeast) and the possible significance of these losses in nutrition of the seed. (Sutcliffe and Pate, 1977). (Reproduced by permission of Academic Press Inc. (London) Ltd)

Mineral	Final amount of element in seeds of the fruit (Wt. per fruit)	Net loss of element during senescence of pod of fruit (wt. per pod)	Loss from pod as a proportion of requirement of seeds (%)
Major elements (mg)			
Potassium	26.7	3.7	13.9
Phosphorus	16.7	2.1	12.6
Magnesium	2.67	0.36	13.5
Calcium	2.35	0.28	11.9
Trace elements (μg)			
Iron	134.5	12.4	9.2
Zinc	67.9	4.5	6.6
Manganese	32.2	3.0	9.3
Copper	16.3	0.84	5.2

2.7.1 Phytic acid

A major proportion of grain phosphorus at maturity is present in phytic acid, the hexaphosphate derivative of *myo*-inositol. Appreciable amounts of calcium and magnesium are similarly bound to form the mixed calcium and magnesium salts of phytic acid (Section 4.2.1).

myo-[^{3}H] Inositol, when fed to ripening rice accumulates in the aleurone grains, suggesting that these are the major sites of phytic acid deposition. Certainly much of the grain phytic acid is found in outer layers of mature grains. This immobilization of essential mineral elements, which may also include iron and zinc, may constitute an additional mechanism for regulating grain development since it removes them from those cellular processes which require them—such as cell division, glycolysis, protein synthesis, and respiration.

Phytic acid is accumulated by many seeds other than cereals, including groundnut, lupin, soybean, and peas. The nutritional significance of this is discussed in Section 4.2.1.

2.8. HORMONES AND SEED FORMATION

A number of hormones are present in plants, including the gibberellins, cytokinins, auxins, and abscisic acid. Although their mechanism of action is not understood in all cases, they are thought to be responsible for the regulation of a wide range of plant processes involved in growth and reproduction. Representatives of the three main classes of hormone as well as abscisic acid have been found to be present in developing pea seeds and cereal grains and there has been much speculation about their role in seed development. Other than recording their presence during the course of seed or grain development, little information is available to suggest how they may control growth and differentiation.

There is, however, some evidence in peas that hormones present in the seeds can affect pod elongation. Certainly this can be loosely correlated with changes in the hormone content of the liquid endosperm of the developing seed (Fig. 2.14).

More dramatically, if seeds are killed early in development by piercing them with a sharp needle through the pod wall, then pod elongation is inhibited. Since developing pea seeds contain relatively large amounts of extractable gibberellins and auxins compared with the pod, it may be that the seed is the source of hormones required for growth of the pod wall. Application of gibberellic acid or another growth-promoting substance, 1-naphthylacetic acid, partially restores elongation of fruits with killed seeds *in vivo*. A mixture of the two results in growth equal to that of the seeded control (Fig. 2.15).

During the later stage of growth of wheat grains there is a dramatic increase (up to 40-fold) in the content of abscisic acid (ABA) per grain. This level remains high from 25 to 40 days after anthesis. Then, in association with natural or forced drying of the grain there is a rapid drop in ABA content. Isolated immature embryos and whole grains were capable of germinating during the mid-growth period, when ABA is still accumulating, but the

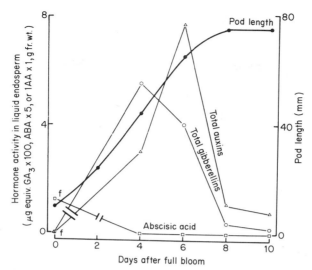

Figure 2.14   Elongation of the pod wall in relation to hormone activity in the liquid endosperm of the Pea seed. f, Flower (Eeuwens and Schwabe 1975). (Reproduced by permission of Oxford University Press)

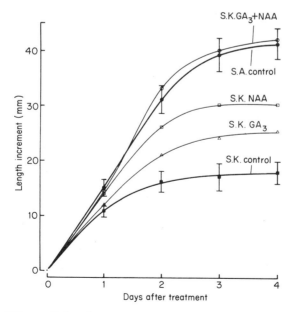

Figure 2.15   Effect of lanolin applications of GA$_3$ (500 ppm) and NAA (1-naphthylacetic acid) (500 ppm) separately and combined, on the elongation of pea fruits with killed seeds *in vivo*. S.K., Seeds killed; S.A., Seeds alive; error bars indicate 95 % confidence limits for mean values (Eeuwens and Schwabe, 1975). (Reproduced by permission of Oxford University Press)

64

germination capacity subsequently declined. As the grain dehydrates, and ABA levels fall, the grain again becomes germinable. Hence, it might be suggested that ABA accumulation prevents precocious germination and premature hydrolysis of starch reserves in the morphologically mature but still unripe grain. Such correlations are hardly satisfactory but unfortunately they form the basis of many of the conclusions on the growth and development of seeds. Other factors are almost certainly involved in the prevention of premature germination, since it has been shown that the ABA content of pea seeds collected at 24 days after full bloom, which germinated, was only slightly lower than that in seeds collected at 16 days, which did not.

REFERENCES

General references

Allison, M. J. (1976). High diastase barley. *ARC Res. Rev.*, 2, 42–44.
Bewley, J. D. and Black, M. (1978). *Physiology and Biochemistry of Seeds*, Springer-Verlag, Berlin.
Brocklehurst, P. A. (1977). Factors controlling grain weight in wheat. *Nature*, 266, 348–349.
Galliard, T. and Mercer, E. I. (1974). *Recent Advances in the Chemistry and Biochemistry of Plant Lipids*, Academic Press, New York.
King, R. W. (1976). Abscisic acid in developing wheat grains and its relationship to grain growth and maturation. *Planta*, 132, 43–51.
Laidman, D. L. and Wyn Jones, R. G. (1979). *Recent Advances in the Biochemistry of Cereals*, Academic Press, London.
Miflin, B. J. and Lea, P. J. (1976). The pathway of nitrogen assimilation in higher plants. *Phytochemistry*, 15, 873–885.
Skarsaune, S. K., Youngs, V. L., and Gilles, K. A. (1970). Changes in wheat lipids during seed maturation. *Cereal Chem.*, 47, 533–544.

Specific references

Beevers, L. (1976). *Nitrogen Metabolism in Plants*, Edward Arnold, London.
Bhatia, C. R. and Rabson, R. (1976). *Science*, 194, 1418–1421.
Brandt, A. (1976). *Cereal Biochem.*, 53, 891–901.
Doll, H., Køie, B. and Eggum, B. O. (1974). *Radiat. Bot.*, 14, 73–80.
Eeuwens, C. J. and Schwabe, W. W. (1975). *J.Exp. Bot.*, 26, 1–14.
Flinn, A. M. and Pate, J. S. (1970). *J.Exp. Bot.*, 21, 71–82.
Nelson, O. E. Personal Communication.
Norton, G. (1978). *Plant Proteins*, Butterworths, London.
Osborne, T. B. (1924). *The Vegetable Proteins*, Longmans, New York.
Perez, C. M., Perdon, A. A., Resurreccion, A. P., Villareal, R. M. and Juliano, B. O. (1975). *Plant Physiol.*, 56, 579–583.
Privett, O. S., Dougherty, K. A., Erdahl, W. L. and Stolyhwo, A. (1973). *J.Am. Oil Chem. Soc.*, 50, 516–520.
Sakri, F. A. K. and Shannon, J. C. (1975). *Plant Physiol.*, 55, 881–889.
Smith, C. G. (1974). *Planta*, 119, 125–142.
Sodek, L. and Da Silva, W. J. (1977). *Plant Physiol.*, 60, 602–605.

Sutcliffe, J. F. and Pate, J. S. (1977). *The Physiology of the Garden Pea*, Academic Press, London.

United States–Canadian Tables of Feed Composition (1969). Publication 1684, National Academy of Sciences, Washington, D.C.

Weber, E. J. (1969). *J.Am. Oil Chem. Soc.*, **46**, 485–488.

Chapter 3

Seed storage and survival

3.1 VIABILITY AND SURVIVAL CURVES

After maturation and harvest, seeds need to be stored until required. The criteria of successful storage depend on the eventual purpose to which the seed will be put. For example, seed for growing must be able to germinate nearly 100% and produce vigorous seedlings in the fields, whereas seed for processing may only be required to be chemically undeteriorated and free from contamination. In practice, the latter requirement appears to go along with maintenance of a reasonably high germinative capacity, although the seeds may not be capable of successful seedling establishment. Seed which can germinate is usually referred to as viable, but viability to the agriculturalist and horticulturist means not only germination but also the ability to form a healthy seedling. In this book the term has essentially the same meaning as germinability. Whatever the seed and whatever the conditions of storage, it is commonly observed that the viability of a batch of seeds, measured as the percentage germinating under standard conditions, remains reasonably static for a while and then begins to decline until eventually none of the seeds will germinate. A great many experiments into the longevity of stored seeds have been carried out and a great deal of variation has been uncovered, but there is general agreement that the critical factors are temperature, moisture content of the seed and oxygen availability (Fig. 3.1). In general, the lower the value for these parameters the longer the seeds remain viable, and it has proved possible on the basis of experiments in sealed containers to devise an equation which successfully predicts the time taken for half the seeds in a batch to die in terms of the storage temperature and moisture content:

$$\log p_{50} = K_v - C_1 m - C_2 t$$

where p_{50} is the stored time in weeks after which half the seeds retain viability, m is the moisture content (%), t is the temperature (°C), and K_v, C_1, and C_2 are constants. This equation has been found to apply to wheat, barley, peas, and broad beans in the ranges 15–25 °C and 11–23% moisture. Inter-species and inter-varietal variations in the value of the constants do occur and in some cases, especially at more extreme levels of temperature and moisture content, the equation fails to predict the correct pattern of the decline in viability. Furthermore, it has been shown that the environmental conditions prevailing during the formation and maturation of the seed can distinctly influence subsequent survival curves. In the case of wheat and barley there may be as much as a 2-fold difference between grain obtained in a good year and that produced in a bad year. A simple point, but one which can be easily overlooked, is that physical damage produced during harvesting or subsequent handling is very likely to influence the retention of viability, usually unfavourably.

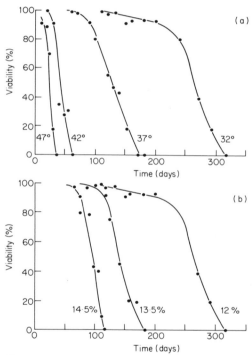

Figure 3.1   Effect of moisture content and storage temperature on seed survival (Roberts, 1961). (a) Rice grains with a moisture content of 12.0% were hermetically sealed in containers with an atmosphere of air and then batches of containers were stored at a range of temperatures. At intervals, containers were withdrawn and the proportion of grains which could still germinate was determined to give the percentage viability. (b) A similar experiment was performed to that described in (a) except that a single storage temperature was used with seeds of different initial moisture contents. (Reproduced by permission of Clarendon Press)

Altogether less work has been carried out on the effect of gases on viability. It does seem, however, that in most cases increases in the partial pressure of oxygen decrease viability, whereas the amount of CO_2 present is not significant. A major reason for the shortage of information on the effect of gases lies in the experimental problems of maintaining a constant atmosphere in a container over a long period of time. Most experiments have been carried out with the seeds stored in containers which were sealed at the beginning of the experiment and opened only when viability was to be tested. Knowledge of the gas composition is thus usually limited to a knowledge of the original gas phase. On the other hand, under most storage conditions marked changes seem unlikely to occur unless the seed sample is contaminated with fungi.

3.2 PRACTICAL ASPECTS OF STORAGE

During storage, several factors other than intrinsic changes in the seeds themselves can reduce the viability of seeds. Seed stores can be attacked by rodents; these animals will consume much of the grain but will also damage and contaminate the remaining material, resulting in an accelerated decline in viability. Prevention of attack by rodents relies very much on the use of rodent-proof stores supported by trapping and poisoning. In a completely different category are insects, mites, bacteria, and fungi, which are all grouped together as they are very dependent on relative humidity and temperature within the grain store. Very little insect activity occurs at temperatures below 17 °C and moisture contents below 8% also prevent insect growth. Mites are much less influenced by temperature and may be active down to 3 °C. However, they are very sensitive to relative humidity and cannot exist below 60%, which is equivalent to about 12% moisture in barley. They require 75% relative humidity before they can indulge in rapid reproduction. The two types of microorganism mentioned represent the most and least dangerous spoilage organisms. Bacteria require 90% relative humidity to survive and so occur only in situations where deterioration is already considerably advanced. Fungi, on the other hand, have relative humidity requirements very similar to those for mites and are also active over a wide range of temperatures. These organisms seem always to be present in harvested seeds and in seed stores, so they represent the most likely mechanism of extraneous spoilage of seeds.

Little will be said about commercial seed storage systems here except that whilst using low moisture contents seems the best way to control spoilage organisms, low temperatures reduce the rate of all chemical reactions and control of both moisture content and temperature represents the best chance of extending seed viability in storage. Control of oxygen does not seem to be worth the problems created. It cannot be overemphasized, however, that when it comes to handling a particular seed in a particular environment, generalizations are not enough, and specific information on the individual situation must be obtained.

3.3 MYCOTOXINS

The spoilage effect of the various types of organism described in the preceding section is usually fairly obvious, e.g. rodents eat the grain and deposit urine and faeces in the store; fungi and bacteria digest the seed material and produce off-flavours. In some cases fungi and bacteria can cause so much heating that the seeds become charred and may even burst into flames. However, fungi have been associated with a different spoilage problem which, because it is not obvious to the eye, is potentially extremely dangerous. This is the production of poisonous compounds, the mycotoxins. A wide variety of fungi have been found to produce toxins at some point in their growth cycle. The commonest genera involved are *Aspergillus*, *Fusarium*, *Penicillium*, and *Claviceps*. The toxins can persist in the seeds long after signs of fungal infection have disappeared and there are cases of maize where the grain was still toxic 12 years after the fungal attack had ceased. Mycotoxins vary in their effects and range of toxicity. Some are relatively specific to one or two species, whilst others appear to be general poisons. Mostly they are hepatotoxins, nephrotoxins, or neurotoxins, but some are carcinogenic, teratogenic, or tremorgenic. Many have distinct antibiotic and antiviral effects. Information on their biological effects is largely restricted to the effects on various animals and relatively little is known about their effects on humans. This is because experiments are not carried out on humans and accidental poisoning is rare as much higher standards are applied to grain going into human food than to that destined for use in animal feed. In recent years, the aflatoxins produced by strains of *Aspergillus flavus* and *A. parasiticus* have attracted much attention. There are at least 18 compounds in this group and they can occur in all of the commonly used seeds, although groundnut meal from tropical countries has been one of the major problem areas. Aflatoxin B_1 is a very dangerous compound; LD_{50} values of 5–7 mg/kg in rats and 0.36 mg/kg in day old ducklings have been reported. This compound is also extremely carcinogenic. It can cause tumours in rainbow trout when present in the food at less than 1 ppb.

Aflatoxin B_1

As mentioned earlier, all the cereals and oilseeds are open to attack by toxin-producing fungi and the conditions for toxin production are similar to those for fungal growth. Seeds with moisture contents in equilibrium with relative humidities of 85% or higher and stored at 25–30 °C are particularly prone to contamination.

Control of mycotoxin contamination rests on the one hand with the prevention of fungal growth by proper handling and drying, and on the other with the identification and separation of infected batches of seed. Lightly infected seed can sometimes still be used in feedstuffs if it is blended and used immediately. More heavily infected seed is not worthless, however, as chemical detoxification is often possible. Ammoniation of aflatoxins in groundnuts, for example, can be used successfully although, of course, the treated seeds are suitable only for non-food, non-feed purposes.

Although storage fungi are the commonest source of mycotoxins, under certain conditions field fungi can also be responsible. For example, grain weathered by exposure to frosts and snow is often infected by *Alternaria* spp. and some of these organisms are mycotoxin producers. A typical example of the kind of compound produced is alternariol.

Alternariol

3.4 INTRINSIC DETERIORATION OF SEEDS

Even under storage conditions where seeds are free from attack by other organisms, viability does eventually decline and finally all of the seeds will die. The length of time that seeds can survive seems to be very variable, stretching from a few days under poor storage conditions to hundreds of years in some cases. The extreme examples quoted of seeds germinating after storage for thousands of years in ancient tombs appear not to be authentic. Seed survival curves depend on both the species and the storage conditions, but the nature of the events which lead to an eventual failure to germinate is still not clear. As the viability of a batch of seeds begins to decline several changes in the more obvious properties of the batch occur. Typically, the seed colour changes a little and the surface loses its natural lustre. The resistance of the deteriorating seeds to stress, such as adverse storage conditions or fungal infection, is reduced compared with the fully viable seeds. The seedlings which are produced show an increasing rate of abnormalities, deformities, and slow growth as viability declines. A good relationship exists between the oxygen uptake of seeds and their germinability; the RQ value (volume of CO_2 produced/volume of O_2 absorbed at N.T.P.) of fully viable seeds is typically low, perhaps 0.6–0.7, but with deterioration the ratio rises to give relatively high values. Most of the above parameters tend to change with, or even after, the decline in viability so there is much interest in finding biochemical changes which occur ahead of changes in viability. One such characteristic is the utilization of glucose by seeds of barley and wheat (Fig. 3.2) rather than the simply uptake of the sugar. In these cases metabolic conversion of glucose had fallen to about 20% of its initial value before there were any effects on percentage germination, shoot growth, or oxygen uptake.

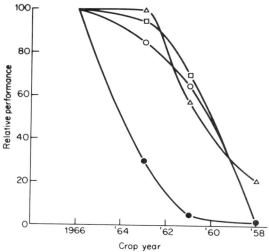

Figure 3.2 Change in vigour parameters during storage of barley (Abdul-Baki, 1969). Samples of barley (Wisconsin X–691–1) were collected from the crops of 1958, 1961, and 1963 and kept in an air-conditioned laboratory maintained at approximately 22 °C. In 1966, these samples, together with grain from the 1966 crop, were tested for utilization of externally supplied glucose (●), amount of shoot growth (○), percentage germination (□), and oxygen uptake (△). The results suggest that on storage the ability to utilize glucose is clearly the first vigour parameter to decline.
(Reproduced by permission of the Crop Science Society of America)

A different approach has been to look for changes in quantity and kind of the major components of the seed. This has not proved to be very successful as changes in carbohydrate, protein, and lipid content are not great. There is a slight decline in starch content due to respiration but this is barely noticeable at lower temperature and moisture contents. Seeds tend to lose non-reducing sugars during deterioration, presumably also due to respiration. A less understood feature of grain is a change in the endosperm which makes the separation of starch by milling much more difficult and the starch yield from heat-damaged grain can be reduced by nearly 50%. The digestibility of proteins also decreases with time and the free amino acid content increases. The latter situation is particularly common in damaged grain. There has been no indication of changes in mineral content but the concentration of vitamins often falls substantially. This has been noticed particularly with thiamine. A good correlation has been observed between an increase in fat acidity and decline in viability, but it seems that this change is due more to lipases secreted by fungi growing on the grain than to intrinsic changes. Fat acidity is usually measured as the amount of potassium hydroxide in milligrams necessary to neutralize the free fatty acids from 100 g (dry weight) of seeds.

Another chemical change which has been noticed is a large loss in phospholipid content when seeds are artificially aged. This suggests that a deterioration in intercellular membrane structure could be responsible for loss of viability. This idea has been supported by studies which have shown that the amount of material leached from seeds is in inverse proportion to the viability of the seeds. Unfortunately, seeds aged under less severe conditions did not show the pronounced decline in phospholipid content, although the amount of leachate increased with deterioration. An alternative explanation of the leaching effect has been put forward stating that, rather than leaching being due to failure of the membrane structure, it reflects the fact that older seeds are less metabolically active and so metabolic intermediates are more easily lost from the cells. It is very important, of course, to take into account any significant physical damage when carrying out the leaching test as large amounts of material may then escape from a relatively small number of seeds.

A number of workers have attempted to find a key enzyme concerned in the decline of seeds in the hope that its assay would provide a simple and accurate indication of seed viability. However, it seems unlikely that any single enzyme could have such an importance, as a decline in metabolic activity is more likely to be a reflection of piecemeal breakdown of the metabolic system and the exact pattern of this breakdown is likely to be different in different seeds and under various conditions. In spite of this, glutamate decarboxylase activity has been used as an indicator of viability in wheat. The function of this enzyme is not understood in plants but its product, γ-aminobutyric acid, is one of the first compounds to be synthesized on imbibition of water by seeds and, in barley at least, the enzyme is converted from an inactive to an active form by molecular oxygen. This information confirms that glutamate decarboxylase is likely to be an important enzyme in germination, but the need for extraction and a relatively complicated assay procedure seem to have restricted its study. In contrast, the metabolic significance of dehydrogenases is well understood and the tetrazolium test, which indicates the extent of general dehydrogenase activity, has proved the most successful biochemical indicator of viability to date. After suitable soaking, which depends on the type of seed under test, the embryo is transferred to a 1% solution of 2,3,5-triphenyltetrazolium chloride buffered between pH 6 and 7. Any absorbed chemical is reduced by active dehydrogenases to an insoluble red formazan so that the embryo becomes stained only where there are active enzymes; dead areas remain clear. Study of the pattern of stained and unstained areas can indicate to a skilled worker whether or not the embryo was viable under a wide range of conditions. The tetrazolium test is not infallible, however, as there are circumstances where apparently viable seeds fail to germinate. It can easily be appreciated that for germination to occur the whole metabolic system must be intact. Even a single lesion at a structural or biochemical level could prevent germination whilst leaving the dehydrogenases, and so the tetrazolium response, intact. Such partial damage or deficiency is likely to occur in old seeds or in seeds which have been exposed to radiation or excessive heat.

Further evidence that a general decline in metabolic activity is a forerunner of failure to germinate comes from observations that the ability to synthesize carbohydrate and protein has often decreased to about half its original value before any noticeable effects on germinability occur.

Another, potentially critical, area in which changes are known to occur during storage is in the chromosomes. As storage proceeds the number of chromosomal abberrations in the embryo increase and eventually become so common that if the embryo germinates chromosomal faults may persist right through into the mature plant, although such errors are often eliminated during the early stages of seedling growth. Whether chromosomal faults are the prime cause of deterioration is not clear. Although such faults accumulate with declining viability it has been pointed out that the number of chromosomal abberrations caused by ionizing radiation is greater rather than lower at lower moisture contents, i.e. the reverse of the effect of moisture content on viability.

3.5 GERMINATION

Germination is a process of change: change from a small quiescent structure living on minimal supplies to an actively growing plant committed to becoming self-sufficient before the reserve materials of the seed are exhausted. The basic conditions for germination are thus the same as those for active metabolism. As seeds almost invariably go through a period of desiccation during maturation, the first phase of germination is the uptake of water (imbibition), although, of course, this can be successful only when the temperature is in an appropriate range. In most cases there is also a need for oxygen, as oxidative respiration is the norm, although in the early stages energy is often released through fermentation of food reserves. In addition to these basic requirements, special factors are needed by some species, e.g. light of a particular wavelength or ethylene produced by soil bacteria. The situation often arises, especially just after harvest, that the sound, healthy, viable seed fails to germinate when provided with the appropriate conditions. This is the phenomenon of dormancy. Eventually dormancy will pass and the seed will become capable of germination. The mechanism of dormancy and the way in which it is lifted are dealt with in Section 3.6 and, for the moment, we can return to consider what happens when a seed germinates.

In most cases the continuous process of germination can be considered as being composed of two major phases:

1. initiation of active metabolism in the embryo, followed rapidly by embryo growth and differentiation, supported by utilization of the immediate embryonic reserve material;
2. continuing growth of the embryo supported by a flow of hydrolysis products from the cotyledons or extra-embryonic food reserve such as the endosperm. This phase continues until the plant is either established as a photosynthetic organism, or dies on exhaustion of the food reserve.

Transition from phase 1 to phase 2 depends on the appearance of a range of hydrolytic enzymes in the food reserve in response to growth of the embryo.

As mentioned above, most seeds are relatively dry tissues so that germination begins with the imbibition of water. The seeds swell owing to the physical adsorption of water by the polymeric reserve material, although this process can occur in both dead and viable seeds. Very rapidly, in the case of healthy seeds, various metabolic changes occur within the embryo. The first to be recognized are increases in the levels of the intermediary metabolites and enzymes associated with energy production, particularly the TCA cycle. On a gross scale this change is reflected in a large increase in the rate of gas exchange, although changes in RQ values depend considerably on the species and the nature of food reserve. Commonly, within a few hours of the uptake of water, protein synthesis is under way, probably using mRNA surviving from the maturation period, rather than freshly transcribed material at the beginning. The next phase involves synthesis of DNA and the beginning of cell division and differentiation of tissues within the embryo. The exact pattern of changes depends very much on the seed involved and the particular conditions of germination; where relevant to the utilization of particular seeds, these changes are described in more detail (Sections 5.2.2, 5.2.3 and 5.8).

3.6 DORMANCY

When mature seeds are collected at harvest time they often fail to germinate if subjected to conditions which normally support germination. This state is known as dormancy but eventually, after a period of after-ripening, the dormancy will pass and the seeds become capable of 100% germination. Occasionally, mature seeds which are capable of full germination will lose this property temporarily, owing to exposure to adverse conditions. This type of dormancy is usually referred to as secondary dormancy. There are several ways of classifying dormancy but none appears to be entirely satisfactory, mainly because the underlying causes of dormancy are still unknown. Furthermore, there does not seem to be a sharp distinction between the need of some seeds for special treatment to induce germination, and seeds which are thought of a dormant but where the dormancy can be broken by chemical or physical treatments. The biological significance of dormancy seems obvious enough as it prevents early germination where the seedling could find itself trying to grow in the winter rather than in the spring. It also tends to spread out the time at which the individual seeds of any one batch eventually germinate so that the chances of adult plants forming are greatly increased.

The direct causes of dormancy are varied and some of the well known factors include immature embryos, a requirement for a short period of light of a specific wavelength, and a requirement for a period at low temperature or fluctuating temperatures. Seeds with hard coats are often dormant and it is common experience that if the seed coat is removed, dormancy is also removed. The reasons for this are not known for certain but it has been suggested that the seed coat could either physically restrict growth of the embryo or act as a barrier to the free exchange of gases or uptake of water. In many cases the presence of specific germination inhibitors has been suspected and it is certainly true that many naturally occurring substances prevent germination. Particularly interesting in this area are several phenols, abscisic acid, coumarin, and auxins at high concentrations. However, it is often difficult to pinpoint the mechanism of dormancy experimentally with any certainty, as the results of a single experiment can usually be interpreted in a variety of ways, e.g. interference with the seed coat could release dormancy through removal of physical restriction of the embryo, through an improvement in permeability of water, gases, or minerals, or through the removal of an inhibitor located either in the coat or in the interior of the seed.

It seems likely that a single cause of dormancy may not exist but it is possible that in many seeds there is a common underlying biochemical explanation. This theory has been derived from a study of the wide range of compounds known to break dormancy in many seeds, e.g. cyanide and hydroxylamine. Surprisingly, in view of the need for O_2 in respiration, the majority of dormancy-breaking compounds actually inhibit oxidative respiration. The suggestion has therefore been made that there is a biochemical imbalance in the dormant embryo which requires some oxidative reaction other than respiration to set the system aright. Studies on the relative evolution of the C_6 and C_1 atoms of glucose as CO_2 have suggested that an increased activity of the pentose phosphate pathway is necessary before germination can begin. This pathway is responsible for the biosynthesis of the pentose sugars which are themselves essential in the biosynthesis of all nucleotides and nucleic acids (Fig. 3.3). The requirement for oxygen is presumably related to the need to oxidize NADPH if the pathway is to act as a net producer of pentoses. It could be that other areas of metabolism are also held in abeyance in the dormant embryo, waiting until a sufficiently high oxygen tension turns the pathway on, and the general idea is biochemically a very attractive one. However, as in nature the key issue would be the penetration of oxygen to the relative parts of the dormant tissues, the normal duration of dormancy is likely to be controlled by factors restricting gaseous movement. This brings us back again into the area of inhibitors, physical permeability barriers, and the influence of hormones. At this level it seems likely that great variation between species will occur even if the basic biochemistry is the same in all plants.

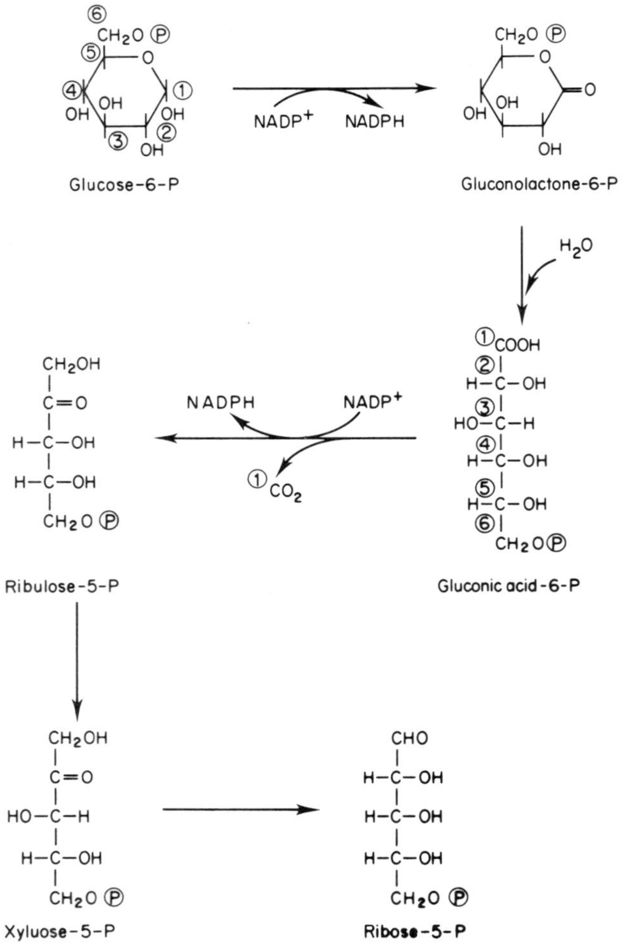

Figure 3.3 Synthesis of pentose phosphates from hexose phosphates. The reactions show the pentose phosphate pathway operating for the synthesis of ribose-5-phosphate, an essential precursor in the synthesis of all nucleotides and hence the nucleic acids and many coenzymes, e.g. NAD+, ATP, and coenzyme A. For each molecule of ribose-5-phosphate produced, two NADPH molecules are also produced, and if this material is not consumed in other biosynthetic pathways then re-oxidation involving molecular oxygen will be necessary to ensure a flow of pentose phosphate. The CO_2 evolved in this pathway comes entirely from the C_1 position of glucose, in contrast to the result of the action of glycolysis where the CO_2 is evolved equally from the C_1 and C_6 positions of glucose. Specifically labelled glucose can therefore be used to determine the relative rates of operation of glycolysis and the pentose phosphate pathway

REFERENCES

General references

Barton, L. V. (1961). *Seed Preservation and Longevity*, Leonard Hill (Books), London.

Beevers, L. (1976). *Nitrogen Metabolism in Plants*, Edward Arnold, London.

Bewley, J. D. and Black, M. (1978). *Physiology and Biochemistry of Seeds*, Vol. 1, Springer-Verlag, Berlin.

Laidman, D. L. and Wyn Jones, R. G. (1979). *Recent Advances in the Biochemistry of Cereals*, Academic Press, London and New York.

Mayer, A. M. and Poljakoff-Mayber, A. (1975). *The Germination of Seeds*, 2nd ed., Pergammon Press, Oxford.

Roberts, E. H. (1972). *Viability of Seeds*, Chapman and Hall, London.

Sharpley, J. M. and Kaplan, A. M. (1976). *Proceedings of the Third International Biodegradation Symposium, Session XIV*, Applied Science Publishers, Barking.

Specific references

Abdul-Baki, A. A. (1969).*Crop Sci.*, **9**, 732–737.

Roberts, E. H. (1961). *Ann. Bot.*, **25**, 381–390.

Chapter 4

Nutritive value of seeds

4.1 DIETARY REQUIREMENTS

Seeds and their by-products form a major part of the diet of many animals. Simple stomached animals, such as man, pigs, and poultry, develop deficiency symptoms on most diets based solely on seeds, since these lack a number of essential nutrients. Ruminant animals, on the other hand, have less rigorous dietary requirements, largely because the microflora of the rumen can synthesize most of the vitamins and all of the essential amino acids required by the animal. However, high costs and the incidence of digestive disturbances associated with diets containing very high levels of seeds mean that supplementation of the seed-based diet is again desirable.

It is clear, too, that different criteria must be used when formulating the dietary requirements of humans and farm animals. In the former case, well-being is the sole criterion, whereas in the latter case, in addition to well-being, the effect of diet on the final composition of the carcass together with the efficiency of food conversion into animal products must also be taken into account.

The primary sources of information on human dietary requirements are published in joint reports of the Food and Agriculture Organization (FAO) and the World Health Organization (WHO) of the United Nations. The figures are expressed as recommended daily intakes (RDI) and are the amount of essential nutrients and energy suitable for a healthy life. These requirements of course vary with age, sex, body weight, activity, and environment of the subject. For example, an adult male student weighing around 65 kg with a moderately active life style requires about 12–13 MJ/day of energy or 0.75 kg of sugar, whereas a female student weighing 55 kg and with a similar life style will require only 9.2 MJ/day (DHSS, 1969).

In addition to energy, other requirements include defined daily quantities of protein, vitamins, and mineral elements. Individual governments, for example those of the USA and UK, publish their own figures which they consider are more suitable to their respective populations and to the particular foods available. Those in the USA are published by the National

Research Council (NRC)–National Academy of Sciences (NAS), and those in the UK by the Department of Health and Social Security (DHSS).

Similarly, nutrient allowances for pigs, poultry, cattle, sheep, and other domestic farm animals are published in the USA by the NRC and in the UK by the Agricultural Research Council (ARC), and cover requirements for maintenance, growth, pregnancy, lactation, and reproduction.

4.1.1 Protein

Proteins in the diet supply the amino acids required for growth of young animals and children; in adults the requirement is for the maintenance of the tissues as well as the additional burden of reproduction, lactation, and more specifically in farm animals for meat, egg, and wool production.

The newborn baby needs about five times as much protein as the adult per unit of weight (Table 4.1). As it grows, so the requirement, on a weight basis, is steadily reduced. A range of nutritional disorders in children is associated with both a qualitative and quantitative lack of protein. These are collectively termed 'protein energy malnutrition' (PEM). If protein only is involved then protein deficiency is manifested in the disease kwashiorkor. However, in the absence of sufficient energy, dietary protein may be used to supply energy rather than amino acids. Thus, protein available for growth is reduced when dietary energy is inadequate. In this case the condition is termed marasmus or, more simply, starvation. Most cases of PEM lie within these two extremes and are generally associated with vitamin and mineral deficiencies as well as disease.

Establishment of nitrogen balance is a technique used to determine the

Table 4.1. Decline in minimum protein requirements with increasing age in healthy humans (Reproduced by permission of the World Health Organisation and the National Academy of Sciences)

Data from FAO/WHO (1973)		Data from NRC (1974)	
Age	Protein requirement*	Age	Protein requirement*
Months		Months	
0–3	2.4	0–6	2.2
3–6	1.85	6–12	2.0
6–9	1.62	Years	
9–11	1.44	1–3	1.76
Years		4–6	1.50
2–5	1.01–1.19	15–18	0.88
10–12	0.74–0.82	23–50	0.80
12–17	0.57–0.78	51+	0.80
Adult	0.55		

*Protein requirement expressed in g protein per kg body weight per day. The figures represent a safe level of intake and, in the case of those from FAO/WHO, include a margin of 30% to allow for individual variation.

minimum protein requirements of an animal. The animal is said to be in N balance when the N output in the urine (U) faeces (F), and scurf (S) equals the N intake (I) in the diet:

$$I = U + F + S$$

If the animal is growing or in production then the N intake must exceed output and the subject is in positive N balance. Conversely, if the N intake is less than that excreted then the subject is in negative N balance. Measurement of N balance then forms the basis of a technique for determining the minimum protein or N requirement of an animal.

Controversy surrounds the estimates of dietary protein required to maintain N balance in humans. Some measure of the differences can be seen in Table 4.1. These figures show that the protein requirements of man, as estimated by FAO/WHO (1973), are less than those accepted in the past and are considerably less than those recommended for citizens of the USA by the NRC (1974). This is partly because the estimates used for obligatory losses of nitrogen from the body were too high. Together with the evidence that some PEM conditions may be relieved by energy rather than protein, the FAO/ WHO (1973) figures suggest that the need for protein may be rather less than previously believed.

Protein requirements of simple stomached animals such as pigs and poultry are determined in feeding trials by measurement of growth rates and carcass quality. The figures are generally expressed for convenience as a percentage of the dry matter of the diet. As with man, the amount of protein required falls with increasing age and weight since relatively more fat is deposited in the later stages of growth (Table 4.2).

Table 4.2. Feeding standards for growing pigs and poultry (McDonald *et al.* 1973). (Reproduced by permission of the Longman group Ltd)

Pigs		Poultry	
Weight (kg)	Crude protein (g/kg dietary dry matter)	Age (weeks)	Crude protein (g/kg dietary dry matter)
9–20	200	0–8	220
20–50	185–200	8–18	175
50–90	150–165	Laying hens	165

The variation in daily protein requirement for growth or maintenance in beef cattle determined on the basis of feeding trials are shown in Fig. 4.1

While proteins can be synthesized in plants *de novo* by the combination of relatively simple nitrogenous precursors such as ammonia with carbon skeletons derived from carbohydrate (Section 2.5.1), this is not the case in animals. Animal protein contains around 22 amino acids and animal cells have lost the ability to synthesize about half of these. Thus animals have a requirement for pre-formed amino acids which can be supplied in the diet

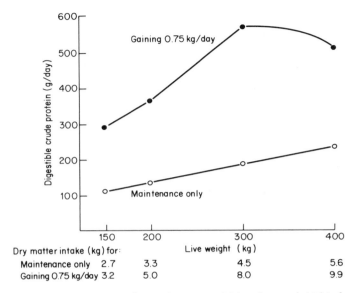

Figure 4.1   Protein requirements for maintenance (O) and growth (●) in beef cattle. Standards are based on the results of feeding trials (McDonald *et al.*, 1973). (Reproduced by permission of the Longman Group Ltd)

either free or in combination as protein. Those amino acids required for normal growth and development in animals are termed the essential amino acids. While ruminants have similar amino acid requirements to monogastric animals, the essential amino acids are derived directly from the biosynthetic activity of rumen microorganisms rather than from the diet. Thus protein quality, in terms of precise requirements for essential amino acids, is rather less important with these animals.

If the protein of the diet is deficient in one or more of the essential amino acids, protein synthesis stops, the nitrogen balance cannot be maintained, and the animal will develop deficiency symptoms. If another protein containing the missing amino acid is added to the diet, nitrogen equilibrium and normal health are restored.

Thus the data for protein requirements (Table 4.2) are of little value unless the protein fed contains adequate amounts of the essential amino acids. Table 4.8 shows the concentration of essential amino acids in the dietary protein that must be present if the requirements for amino acids are to be met. Standards for essential amino acid requirements of chicks and pigs have been devised, but these are only a guideline since requirements are influenced by interactions between the various amino acids and also between them and other nutrients. For example, cystine can be synthesized from methionine, thus effectively lowering the absolute dietary requirement for cystine. In chicks the requirement for glycine is increased when dietary levels of methionine or arginine are low.

In practice, the inefficient utilization of dietary protein occurs because only a few amino acids are limiting. These are methionine, lysine, and arginine in the case of chicks; for pigs only lysine is likely to be limiting.

A number of methods have been devised for measuring the nutritive value of proteins, none of which is entirely satisfactory. The ideal technique would be to determine the essential amino acid content of the protein and to assess nutritive value in terms of the known amino acid requirements of the particular animal. However, there are experimental difficulties in the routine assay of such essential amino acids as tryptophan, cystine, and methionine. In practice, problems with amino acid imbalance, antagonism, and even toxicity mean that measures of protein quality based on amino acid composition do not always correlate well with results from direct feeding experiments. A commonly quoted measure of protein quality is biological value. This is a direct measure of the dietary protein that can be utilized by the animal for the synthesis of body tissues and metabolites, and is defined as the percentage of dietary nitrogen absorbed which is retained by the animal. It is determined from nitrogen balance trials in which the animals are fed known amounts of the protein to be evaluated. The nitrogen excreted, which is derived originally from the dietary protein, is then measured. The biological value of the protein can then be calculated from these results. The technique has a number of disadvantages. One is that, as animals differ in their amino acid requirements, no single value for biological value can be assigned to each protein. Another is that, since the biological value depends on the relative amounts and range of essential amino acids in a given protein, a protein with a deficiency of any essential amino acid will have a low biological value. If this protein is then supplemented with another which is relatively rich in the same amino acid (but not excessively so), the combined proteins will have a higher biological value than the original protein. Thus, biological value cannot be used to predict the quality of a protein when used as a supplement to a given diet.

In practice, dietary protein for simple stomached animals, including man, is evaluated both in terms of its quantity and also in terms of the contents of those essential amino acids most likely to be deficient.

The quantity of protein in a diet is generally quoted as 'crude protein' in g per kg dietary dry matter. This is calculated from the total nitrogen content of the diet (Section 4.2.3), assuming that protein contains 16% nitrogen. That is, if the nitrogen content is multiplied by 100/16, or 6.25, an approximate value for protein is obtained. The value includes all nitrogenous substances as well as protein. Hence it is called 'crude' rather than 'true' protein.

4.1.2 Energy

Animals require dietary energy for muscular activity and, by providing energy for the synthesis of essential metabolites, for the maintenance of the tissues. Energy supplied in excess of maintenance may be utilized for production and, in younger animals, for growth. The energy is contained in the major constituents of the diet, carbohydrate, protein, and lipid with carbohydrate containing the least energy and lipid the most, on a weight basis. Carbohydrate is the major source of dietary energy, and certainly the cheapest. The energy requirements of animals and man vary markedly with age, occupation, climate, environment, weight, and sex (Table 4.3).

Table 4.3. Recommended daily intakes of energy for the UK, 1969 (DHSS, 1969). (Reproduced by permission of the Controller of H.M. Stationery Office)

Age range (years)	Occupational category	Body weight (kg)	Energy requirement (MJ)
Boys: 9–12	—	31.9	10.5
Girls: 9–12	—	33.0	9.6
Men: 35–65	Sedentary	65	10.9
	Very active	65	15.1
Women: 18–55	Most occupations	55	9.2
	Very active	55	10.5
75 and over	Sedentary	53	8.0

The gross energy of a food (i.e. the quantity of energy as heat resulting from the complete oxidation of a known weight of food) cannot be used as the measure of energy available to the animal since it does not include any allowance for energy losses during digestion and metabolism. For example, cellulose and hemicellulose are digested to only a small extent by simple stomached animals, including man, and even in ruminants the normal diet is fibrous and of variable digestibility, i.e. 60–80%. The commonest energy standard is metabolizable energy, which is the gross energy of a food less the energy lost in faeces, urine and, if a ruminant, in the combustible gases (mainly methane) derived from the digestive tract. Much energy can also be lost as heat, particularly after a meal, owing largely to the inefficiency with which the absorbed nutrients are metabolized. Subtraction of this from the metabolizable energy gives the net energy of a food, i.e. that energy actually available for maintenance and production. Although the metabolizable energy is relatively constant for each species of animal, the efficiency with which it is used varies. Hence the net energy of a food will also vary, making its determination difficult and feeding systems based on its use highly complex. Nevertheless, net energy systems are probably the most satisfactory for assessing the value of foods for ruminant animals. In practice, however, metabolizable energy is probably the most quoted energy standard, particularly for pigs and poultry.

4.1.3 Vitamins

Vitamins are organic substances required in small amounts for normal growth and maintenance of animal life. Since they cannot be synthesized by

Table 4.4. The vitamins and their functions

Solubility	Vitamin	Chemical name	Function
Water soluble	B_1	Thiamine	Part of the coenzyme thiamine pyrophosphate: concerned in carbohydrate, fat and protein oxidation
	B_2	Riboflavin	A constituent of the flavoproteins. Similar function to thiamine
	—	Nicotinamide	A constituent of the coenzyme NAD. Similar function to thiamine
	B_6	Pyridoxine	Constituent of coenzyme pyridoxal phosphate: concerned in glycogen mobilization and amino acid metabolism
	—	Pantothenic acid	Constituent of coenzyme A. Similar function to thiamine
	—	Biotin	Concerned in fatty acid synthesis
	—	Folic acid	Involved in amino acid interconversions
	B_{12}	Cyanocobalamin	Concerned in blood glucose synthesis in ruminants; blood formation
	C	Ascorbic acid	Exact function uncertain
Fat soluble	A	Retinol	Vision; bone development; integrity of cell membranes
	D_2 D_3	Ergocalciferol Cholecalciferol	Required for calcium absorption and mobilization; also for bone formation and growth
	E	tocopherols	Exact biological function unknown; functions as an antioxidant
	K_1	Phylloquinone	Required for blood clotting

the body in sufficient amounts they must be supplied in the diet. Ruminants are an exception since many of the vitamins can be synthesized by rumen microorganisms and supplied directly to the animal after absorption in the digestive tract.

The vitamins are most conveniently classified into two main groups: those which are water soluble and those which are fat soluble. The functions of the vitamins are many and varied, as shown in Table 4.4. All those listed are probably required in the human diet. Cattle and sheep, on the other hand, require only the fat soluble vitamins A, D, and E. So far as is known only man and other primates, guinea pigs, the red vented bulbul bird, and the fruit-eating bat require a dietary source of vitamin C. Some vitamins exist as provitamins, which can be readily converted into the vitamin by the animal. For example, a number of carotenoids are precursors of vitamin A, and two sterols, ergosterol and 7-dehydrocholesterol, are precursors of vitamins D_2

and D$_3$, respectively. The vitamin requirements of a range of animals are shown in Table 4.5.

Table 4.5. (Vitamin requirements of animals and man.) Figures for chicks and pigs expressed as amount per kg of food. Figures for men and women are recommended daily requirements.

Solubility	Vitamin (mg)	Chicks* 0–4 weeks	Pigs† Weight 0–100 kg	Men‡ 19–22 years, Weight 67 kg	Women§ 19–22 years, weight 58 kg
Water Soluble	Thiamine	1.0	1.5	1.5	1.1
	Riboflavin	4.0	2.5	1.8	1.4
	Nicotinic acid	28.0	20.0	20.0	14.0
	Pantothenic acid	10.0	10.0		
	Pyridoxine	3.5	1–2.5	2.0	2.0
	Vitamin B$_{12}$	0.02	10–18	3.0	3.0
	Choline	1,300.0	850		
	Vitamin C	Not normally required	Not established	45.0	45.0
	Folic acid	1.5	Not established	0.4	0.4
	Biotin	0.15	Not established		
Fat soluble	Vitamin A (μg)	396.0	417.0	1 000.0	800.0
	Vitamin D (μg)	10.0	5.0	10.0	10.0
	Vitamin E (mg)	10.0	6.7	10.0	8.0
	Vitamin K (mg)	0.5	0.06§		
	Linoleic acid (g)	10.0			

*Data from ARC (1975). (Reproduced by permission of The Agricultural Research Council.)
†Data from Whittemore and Elsley (1977). (Reproduced by permission of Farming Press Ltd.)
‡Data from NRC (1974). (Reproduced by permission of the National Academy of Sciences.)
§Baby pigs.

4.1.4 Minerals

As with plants, animals require a range of essential mineral elements for normal growth and development. These, together with their functions, are listed in Table 4.6. They are divided into two groups, major and trace, depending on their concentration in the body. A number of other elements, such as selenium, vanadium, nickel, tin, and silicon, are of nutritional importance but can only loosely be described as essential. This is because an experimental deficiency of any one of them has one common result, viz. suboptimal growth. Normal growth can be restored by trace supplementation. The mineral requirements for a number of animal species are shown in Table 4.7 together with the mineral composition of barley. Should these requirements not be met, then a deficiency disease can result. For example, calcium deficiency in young animals can result in rickets, a deformation of the bones. Lack of copper can result in a variety of symptoms, many of which are associated with degeneration of the nervous system.

4.2 NUTRITIVE VALUE OF CEREAL GRAINS

The nutritive value of a seed is a measure of how nearly it can supply the

Table 4.6. The essential mineral elements and their functions

Mineral	Main biological functions
Major elements	
Calcium	Major constituent of bones and teeth
Phosphorus	Major constituent of bones and teeth
Potassium	Osmotic regulation of body fluids
Sodium	Osmotic regulation of body fluids
Chlorine	Involved in gastric digestion; osmotic regulation
Sulphur	Constituent of some essential amino acids and vitamins
Magnesium	Required for protein synthesis and enzyme activation
Trace elements	
Iron	Constituent of haemoglobin and cytochromes
Zinc	Constituent of some enzymes
Copper	Contained in cytochrome oxidase. Required for red blood cell formation.
Manganese	Required for bone formation; enzyme activation
Iodine	Required for thyroid activity
Cobalt	Constituent of vitamin B_{12}
Molybdenum	Constituent of xanthine oxidase
Selenium	Involved with vitamin E in prevention of lipid oxidation
Chromium	Involved in glucose oxidation

Table 4.7. Mineral requirements of animals. Amounts present in barley are shown for comparison. Figures for chicks and pigs expressed as amount per kg of food. Figures for women are recommended daily requirements.

Mineral	Sheep* weight 40 kg	Chicks† 0–4 weeks	Pigs‡	Women§ 19–22 years, weight 58 kg	Concentration in barley (amount per kg)
Major elements (g)					
Calcium	6.8	12.0	4–10	0.8	0.5–0.8
Phosphorus	3.4	6.0	4.8	0.8	4.0
Potassium		2.5	$\cong$2.5		4.5
Sodium	0.96	1.5	0.2–1.2		0.2
Chloride		1.4	0.3–1.8		
Magnesium	0.82	0.4	0.3–0.8	0.3	1.3
Trace elements (mg)					
Iron	30	75.0	$\cong$60	18	54.0
Zinc	35	40.0	40–100	15	16.0
Copper	5	4.0	3–10		7.8
Manganese	40	50.0	5–40		16.0
Iodine	0.12	0.4	Some	0.1	
Cobalt	0.12		Some		
Selenium	0.10				0.10

*Data from Scottish Agricultural Colleges (1978). (Reproduced by permission of the Scottish Agricultural Colleges.) Major elements as total daily requirements. Trace elements in mg/kg of food. daily day matter intake 1.44 kg. Daily live weight gain 200 g.
†Data from ARC (1975). (Reproduced by permission of the Agricultural Research Council.)
‡Data from Whittemore and Elsley (1977). (Reproduced by permission of Farming Press Ltd.)
§Data from NRC (1974). (Reproduced by permission of National Academy of Sciences.)

nutrient requirements of the animal. Thus, the amounts of energy, minerals, vitamins and protein quantity and quality should be specified, together with their availability to the animal. Nutritive value is also adversely affected, particularly in oilseeds, by the presence of toxic substances. Hence these should always be considered in any assessment of seeds for feeding. Again, the fibre fraction, consisting largely of cellulose and hemicellulose, although generally unavailable to man and other simple stomached animals, may still be beneficial to those animals, since, by accelerating the rate of passage of digesta, this fraction is considered to lower the incidence of intestinal disorders.

On the whole, the composition of cereal grains is remarkably constant. The major causes of variation are in the water content, which may be between 10 and 14%, and in the protein content, which in wheat may be in the range 6–22%. In general, cereal grains contain 60–70% starch, 8–12% protein, 1–3% mineral elements, 2–5% lipid, 12–15% water, and 3–11% fibre.

Cereal grains are regarded as a major source of energy for most domestic animals and man. They do not, however, provide a complete diet, particularly in the case of simple stomached animals. This is mainly as a result of deficiencies in certain of the essential amino acids, notably lysine, methionine, and tryptophan.

4.2.1 Minerals

If the mineral requirements for pigs and chicks are compared with the composition of a representative cereal such as barley, then it can be seen that this is seriously deficient in calcium and only rather less so for phosphorus (Table 4.7). Furthermore, although most of the essential mineral ions are present in cereal grains, many are not in a form suitable for absorption by the animal's digestive system. For example, much of the calcium, phosphorus, and magnesium is present in combination with phytic acid.

Experiments with chicks have shown that the phosphorus of calcium phytate is utilized only 10% as effectively as disodium phosphate; in laying hens, however, the phytate phosphorus is used about half as well. In sheep, the hydrolysis of calcium and magnesium phytates is catalysed by bacterial phytases present in the rumen to give free calcium and magnesium phosphates, which can be absorbed, and myo-inositol:

Calcium phytate   Myo-inositol

In man it appears that the ability to use phytate may vary, since in India, which is probably typical in this respect of Asia and Africa, despite low levels of calcium in the diet, bones and teeth calcify normally. It may be that adaptation to a low calcium diet may result in an increased phytase secretion in the digestive tract. The phytates and much of the total grain calcium are in the outer layers of the grain and so any products in which these are removed, such as low extraction flours or polished rice, will be relatively lower in these compared with the original intact grain. On the other hand, the availability of those elements which form complexes with phytic acid is increased in leavened bread since yeast contains an active phytase. This is particularly clear in the case of iron, which also forms insoluble complexes with phytic acid and in any case is slightly deficient in cereal grains, since in spite of a high iron intake, anaemia is prevalent within certain groups of people in Iran whose staple food is unleavened wholemeal bread. It is assumed that in the absence of phytase activity most of the iron is made unavailable by combination with phytic acid. Fortunately, the adult's need for iron is low since, although little is absorbed, little is excreted. Women, of course, require more iron than men since there are monthly losses in menstruation. Pregnancy, too, makes heavy demands for extra iron.

Of the remaining mineral elements, zinc and manganese may also be slightly deficient.

4.2.2 Protein and amino acids

Cereal grains generally contain around 10% protein (on a dry matter basis), although individual varieties can occasionally contain much more. For example, some wheats have as much as 22% protein. Maize and rice have the lowest protein contents, often around 8–9%, and oats and wheat with 12–13% are higher in protein content than barley and sorghum.

Cereal protein is of poor quality and for most simple stomached animals should be fed supplemented with protein from animal sources. Wheat and maize proteins are notably poor in lysine. On the other hand oats (Table 4.8) has a larger amount of lysine and this might explain why the value of cereal proteins for promoting growth in young chicks is in the order oats > barley > maize or wheat. As we shall see (Table 4.16), soybean meal is particularly high in lysine and will therefore complement cereals very satisfactorily. Of course, in theory the total amino acid requirement of, say, a growing pig can be met by feeding barley alone, but the extra carbohydrate eaten would make the animal too fat, and all the non-limiting amino acids, which would include a number of essential amino acids, would be lost.

The quality of maize protein is particularly poor and this is reflected in its low biological value (Table 4.9). In those countries where maize is grown and animal production is a major part of the agriculture, most is fed directly to farm animals. However, it has such favourable agronomic characteristics, notably high yield, that in many countries it is used directly for human consumption. The poor quality is due to the very low lysine and tryptophan content of zein, which accounts for about 50% of the total protein of the maize grain.

Table 4.8. Amino acid dietary requirements—how far are they met by oat and wheat proteins?

Essential amino acids	Amino acid requirements (mg/g protein) in diet			Amino acid composition (mg/g protein)		
	Growing pigs*	Turkey poults†	Infants‡	Whole wheat flour§	Oats¶	Human milk#
Histidine	15	17.8	14	20.8	21	26
Isoleucine	35	32.1	35	33.6	36	46
Leucine	50	50	80	67.2	71	93
Lysine	55	46.4	52	24.0	38	66
Methionine + cystine	31	28.5	29	41.6	39	42
Phenylalanine + tyrosine	35	50	63	75.2	84	72
Threonine	32	32.1	44	27.2	33	43
Tryptophan	10	7.8	8.5	11.2	14	17
Valine	35	35.7	47	44.8	50	55

*Data from Whittemore and Elsley (1977). (Reproduced by permission of Farming Press Ltd.)
†Data from ARC (1975). (Reproduced by permission of the Agricultural Research Council)
‡Based on a safe level of intake of 2 g of protein per kg body weight per day. Data from FAO/WHO (1973). (Reproduced by permission of the World Health Organization.)
§Data from McMance and Widdowson, (1978). (Reproduced by permission of the Controller of H.M. Stationery Office.)
¶Data from Harvey (1970). (Reproduced by permission of the Commonwealth Bureau of Nutrition.)
#Data from FAO/WHO (1973). (Reproduced by permission of the World Health Organization.)

Table 4.9. Biological values of cereals. (Eggum, 1973). (Reproduced by permission of the International Atomic Energy Agency)

	Biological	value
Cereal grain	Rats	Baby pigs
Barley	71.8	80.8
Maize	58.1	72.6
Oats	70.4	76.4
Rye	76.7	79.7
Sorghum	52.2	73.5
Wheat	59.0	71.2
Milk	—	95–97

From results with children living in German orphanages in the period 1947–49, Widdowson and McCance (1954) found wheat to be a good source of protein, requiring only minimal supplementation with animal protein (8 g/day).

After lysine, the next limiting amino acids are methionine and probably threonine. In maize diets tryptophan can also be limiting. Of course, it is only relevant to consider individual amino acids in the context of the total amino acid supply in mixed diets, and particularly where animal protein is included amino acid deficiencies are rare.

When the amino acid requirements of animals (Table 4.8) are compared with the amino acids actually present in cereal grain, it can be seen that, with the exception of lysine and possibly threonine, barley protein meets the demands of turkeys and pigs. Where the requirements of infants are considered, both wheat and barley are less satisfactory. The equivalent figures for human milk, generally considered to have the ideal amino acid composition for infants, are shown for comparison.

Wheat is a major part of the human diet and is normally eaten after processing into flour. On the whole the methods used either in processing or in cooking cause significant nutritional damage by denaturing the protein and lowering its digestibility.

The property of elasticity, characteristic of wheat glutens, which makes wheat flour suitable for breadmaking, can lead to detrimental effects when wheat is fed uncooked to animals. A pasty mess is formed in the mouth, particularly if the grain is finely milled, and digestive disturbances can ensue. Thus it is ill-advised to feed wheat in quantity to pigs or poultry.

4.2.3 Energy

Most of the energy in any diet suitable for monogastric animals (including man) comes from starchy carbohydrates, the remainder being derived from protein and, to a much smaller extent, fat. The crude fibre fraction, which includes hemicelluloses, celluloses, and lignin, provides energy for the ruminant animal but, in the case of the monogastric animal, because it is so poorly digested, it can actually reduce the energy value of the diet.

In formulating nutritive values for farm animals, the starch content is not quoted. Instead, the figures for metabolizable energy are considered to be more relevant, together with the figures for lipid (ether extract), crude fibre and crude protein.

The lipid content is determined by extraction of the food with light petroleum. The soluble material is termed the ether extract or lipid fraction and contains fatty acids and triglycerides as well as many other compounds including waxes, steroids, and some pigments. The crude fibre content is determined by treating the residue, after ether extraction, with boiling alkali and acid. The organic residue is defined as the crude fibre and contains cellulose, hemicelluloses, and lignin. The total nitrogen in the diet, from which the crude protein content can be calculated (Section 4.1.1), is determined by a modification of the Kjeldahl sulphuric acid digestion technique. During the acid digestion all of the nitrogen present, with the exception of that in nitrate and nitrite, is converted to ammonia. This can be distilled off after the addition of alkali and determined by titration with standard acid.

Of the grains as harvested (Table 4.10), oats have the lowest metabolizable energy and maize the highest. Overall, however, the digestibility of cereal lipid, protein and starchy carbohydrate is generally high. The high crude fibre of oats (12%) probably accounts for their low metabolizable energy, although it is clear that there are species differences in the digestibility of the different starches (Section 4.2.4). Certainly the overall digestibility of dietary energy is high in the absence of fibre. This is particularly clear in the case of maize and sorghum, which have very low fibre contents and a gross energy digestibility of 90% in pigs, compared with 67% for oats. Thus oats is unpopular for monogastric animals. Seed is generally sold unprocessed for ruminant animals or dehusked as the basis of several breakfast cereals.

Table 4.10. Gross and metabolizable energy of cereals: energy-containing components (MAFF, 1975). (Reproduced by permission of the Controller of H.M. Stationery Office.) All figures are on a dry matter basis.

Cereal	Dry matter content (g/kg)	Gross energy (MJ/kg)	Metabolizable energy (MJ/kg)*	Crude fibre (g/kg)	Ether extract (g/kg)	Crude protein (g/kg)
Barley	860	18.3	13.7	53	17	108
Sorghum	860	18.8	13.4	21	43	108
Maize	860	19.0	14.2	24	42	98
Millet	860	18.7	11.3	93	44	121
Oats	860	19.0	11.5	121	49	109
Rice (polished)	860	18.0	15.0	17	5	77
Wheat	860	18.4	14.0	26	19	124

*Ruminants

Energy digestibility is also improved by reduction in the particle size of the grains by grinding. This process may also damage the starch granules and increase their digestibility (Section 4.2.4). The difference is greatest between intact grains and coarse particles. Little improvement can be achieved by fine grinding (Fig. 4.2).

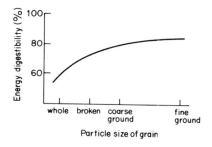

Figure 4.2 Improvement in energy digestibility by reduction in particle size of grains fed to pigs (Whittemore and Elsley, 1977). (Reproduced by permission of Farming Press Ltd)

4.2.4 Starch digestibility

The susceptibility of starches from different plant sources to attack by digestive enzymes varies markedly (Fig. 4.3). For example, in maize the sugary 2 gene is outstanding in that it induces a low temperature of starch gelatinization (58 °C) compared with that of normal maize starch (70 °C). Starch from sugary 2 mutants is also highly susceptible to enzymatic digestion. This may be related to the fact that the homozygous sugary 2 gene induces granules with extensive internal splits, similarly to wrinkled pea starch—except that the granules are much smaller. On the other hand, amylose-extender starch, which has peculiarly shaped long bulbous granules, has a high gelatinization temperature (92 °C) and is high in amylose (60–70%), is very resistant to the action of digestive enzymes.

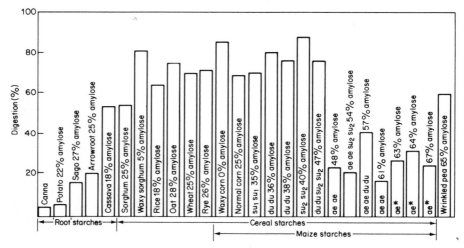

Figure 4.3 Comparison of the susceptibility of various starches to pancreatic digestion. Asterisk denotes unknown composition (Sandstedt, 1965). (Reproduced by permission of the American Association of Cereal Chemists)

The differences in susceptibility are then apparently not related to amylose content, gelatinization temperature, or the ability of the granule to absorb cold water. It is clear, however, that damaged starch is more readily digested.

4.2.5 Processed cereal products

The cereals eaten by man are generally processed. These foods are mainly derived from wheat, oats, maize, rice, and various millets. The nutritive value and energy contents of some of these processed foods are shown in Table 4.11. It is clear that a number of changes in nutritive value take place on cooking or processing. This is, in the main, a result of changes in the water content where, for example, maize as harvested has a water content of around 14% but cornflakes have only 3%. Similarly, wholemeal bread has

40% water compared with the 14% of the intact grain as harvested. Thus, the energy value of the products varies from 5.2 MJ/kg for boiled rice to 15.7 MJ/kg for cornflakes on a fresh basis. It is clear from the figures in Table 4.11 that white flour has a higher energy content than wholemeal flour and that this difference is maintained in the bread made from these flours. This is due to the fact that white flour contains more available carbohydrate in the form of starch and is in spite of the fact that white bread contains less protein and less fat. It must be noted, however, that only about half the fat in wholemeal flour is digested and absorbed by man. The higher fat content of the wholemeal flour is due to the presence of the embryo and aleurone, both of which are rich in lipid.

Table 4.11. Energy value of processed cereal products: energy-containing components (McCance and Widdowson, 1978). (Reproduced by permission of the Controller of H.M. Stationery Office) Figures expressed per kg fresh food.

Product	Dry matter (g/kg)	Starch and dextrins (g/kg)	Dietary fibre (g/kg)*	Crude protein (g/kg)	Energy value (g/kg)†	Lipid (g/kg)
Wheat flour, white (72% extr.)	855	733	30	113	14.3	12
Wheat flour, wholemeal (100% extr.)	860	635	96	132	13.5	20
Bread, wholemeal (100%extr.)	600	397	85	88	9.2	27
Bread, white	610	479	27	78	9.9	17
Rice, polished	883	868	24	65	15.4	10
Rice, boiled	301	296	8	22	5.2	3
Chapatis, made with fat	715	465	37	81	14.1	128
Cornflakes (Kelloggs)	970	777	110	86	15.7	16

*Values not comparable with crude fibre. They include pectic substances as well as hemicellulose, cellulose, and lignin.
†Calculated from amounts of protein, lipid and carbohydrate in the foods using energy conversion factors. The figures are an estimate of metabolizable energy.

It is interesting that if the average British woman were to eat sufficient wholemeal flour (equivalent to 1½ loaves per day) to meet her energy requirements, she would also obtain more than enough protein, iron, and thiamine (see Table 4.12). In theory she would also obtain an adequate supply of the essential amino acids.

Table 4.12. Percentage of recommended intakes of some nutrients met by sufficient flour to satisfy energy requirements of UK women (DHSS, 1969). (Reproduced by permission of the Controller of H.M. Stationery Office)

Flour	Amount per day required (g)	Energy	Protein	Ca	Fe	Thiamine	Riboflavin
Wholemeal	700	100	156	52	199	307	52
White (fortified)	633	100	137	187	110	221	14

4.2.6 Vitamins

The cereals are markedly deficient in a number of vitamins: they contain no vitamin B_{12} or vitamin C and are practically devoid of vitamin A, D, and K activity (Table 4.13). When these figures are compared with animal requirements (Table 4.5) it can be seen that the levels of thiamine and pyridoxine in the grains as harvested are generally adequate. Thiamine levels, however, are low in rice. Barley and maize are relatively deficient in pantothenic acid, but oats, sorghum, and wheat have satisfactory levels. More serious are the low levels of riboflavin in all cereals. Indeed, the human adult, if existing solely on a cereal diet, would have to eat around 1 kg of maize as harvested in order to satisfy the daily requirement. Most of the cereals, with the exception of maize, are deficient in carotene, the percursor of vitamin A; around 150 g of sorghum contains the daily requirement, compared with 30 g of maize.

Table 4.13. Typical vitamin contents of cereals (McDonald *et al.* 1973). Reproduced by permission of the Longman Group Ltd). Figures expressed in mg/kg fresh weight. Dry matter content 870 g/kg.

Vitamin	Barley	Maize	Oats	Rice	Sorghum	Wheat
Carotene	0.4	4.0	—	—	1.2	—
Vitamin E	6.2	0.4	6.0	7.2	—	15.8
Thiamine (B_1)	5.2	4.2	6.4	2.8	4.2	5.0
Riboflavin (B_2)	2.0	1.4	1.6	0.6	1.4	1.2
Nicotinic acid	60.0	22	16.0	36.0	44.0	58.0
Pantothenic acid	6.6	5.6	13.2	6.4	11.4	12.2
Pyridoxine (B_6)	3.0	7.6	1.2	6.4	5.4	4.8
Choline	1 050.0	570.0	1 100.0	920.0	700.0	850.0

Since many of the vitamins are concentrated in the aleurone layer and the embryo, any processing which removes the outside layers will lower the vitamin concentration in the remainder. This is particularly the case with the B vitamins (Table 4.14), whose concentrations are significantly lower in white flour (70% extraction) than in wholemeal flour (100% extraction).

Table 4.14. Vitamin composition of 100% and 72% extraction flours (McCance and Widdowson, 1978). (Reproduced by permission of the Controller of H.M. Stationery Office). Figures expressed in mg/kg flour.

Vitamin	100% extr.	72% extr.
Thiamine	4.6	3.4
Riboflavin	0.8	0.3
Nicotinic acid	56.0	20.0

4.2.7 Lipids

The lipid content of cereal grains varies (see Table 4.10), being poorest in maize and highest in oats. Again, since much of the lipid is in the embryo and aleurone, any processing which removes the outer layers also removes most of the lipid. The major nutritional importance of cereal lipid is its contribution to the energy requirement, but it also can have a deleterious effect on the keeping quality of the carcasses of monogastric animals, particularly pigs. This is because cereal lipids are unsaturated (see Chapter 2) and tend to produce soft body fat. This effect is greatest when animals are fed on oats or maize, which have the highest lipid contents of the cereal grains.

4.3 NUTRITIVE VALUE OF OILSEEDS AND THEIR BY-PRODUCTS

The greatest economic interest in oilseeds is as a source of cooking oils. The residues after oil extraction, which are cakes or meals rich in protein, are, with the possible exception of the soybean, generally considered as by-products and used as feed for farm animals. Additionally, groundnuts, coconuts, sunflower and sesame seeds can all be eaten directly by man, although sunflower and sesame are mainly cooked. Soybeans are considered unpalatable unless cooked. Other oilseeds, such as rapeseed, cotton seed and palm kernels, are not eaten directly by either animals or man.

The oilseeds vary greatly in protein content and, as a consequence, vary also in oil content (Table 5.3). In general, the higher the oil content, the lower is the protein content. Thus soybeans, with their low oil content, contain as much as 40% protein.

In this section we shall consider mainly the oilseed cakes, meals, and flours remaining after removal of the oil rather than the seeds as harvested, as was done in the case of the cereals.

The methods of processing may adversely affect the nutritive value of the seeds (Chapter 5), particularly the protein. In spite of this the protein content and quality of oilseed cakes, meals, and flours is much superior to those of the cereals (Tables 4.15 and 4.16). The protein content of oilseed meals is enormously variable (Table 4.15). There is, of course, considerable natural variation between the different oilseeds but, in addition, decortication or removal of the seed husk has a large effect on final protein content. For example, in sunflower cake the protein content is doubled if seeds which have been first decorticated are used. Few meals based on undecorticated oilseeds are now sold. The oil content of the various meals, as measured by the amount extracted in ether, is also affected by the method of processing used. Solvent extraction results in meals of much lower oil content (Table 4.15) than those which have been subject to the expeller screw extraction process (Section 5.14).

Most of the seeds from which the meals are derived contain a number of toxic substances which frequently pass through to the products. One result is that oilseed by-products are less popular for feeding than they might otherwise be.

Table 4.15. Protein and oil content of oilseed cakes and meals (MAFF, 1975). (Reproduced by permission of the Controller of H.M. Stationery Office). Figures expressed per kg dry matter.

Oilseed meal of cake	Crude protein (g/kg)	Ether extract (g/kg)
Coconut cake	236	81
Cottonseed cake, Brazilian	304	61
Groundnut meal, decorticated and solvent extracted	552	8
Palm kernel cake	216	68
Palm kernel meal, solvent extracted	227	10
Soybean cake	504	66
Soybean meal, solvent extracted	503	17
Sunflower cake, decorticated	413	152
Sunflower cake, undecorticated	206	80

Table 4.16. Mineral ions in oilseed meals (NAS, 1969). (Reproduced by permission of the National Academy of Sciences). Figures expressed per kg dry matter, after solvent extraction

Mineral	Soybean meal (g/kg)	Cottonseed meal (g/kg)
Calcium	3.6	1.7
Phosphorus	7.5	13.1
Magnesium	3.0	6.1
Sodium	3.8	0.4
Iron	0.13	0.33
Manganese	0.031	0.023
Copper	0.041	0.021
Cobalt	0.0001	0.00016

4.3.1 Minerals

When compared with actual mammalian requirements the mineral ion content of the oilseeds (Table 4.16) is, with the exception of calcium, adequate. However, it is likely that the relatively high phosphorus content will reduce available calcium, which in any case is deficient. Furthermore, phytic acid, which is present in all seeds (Section 4.2.1), will reduce the availability of calcium, magnesium, and phosphorus to the animal.

4.3.2 Protein and amino acids

On the whole, oilseed protein contents are three to four times those found in cereal grains. Hence oilseed meals are often referred to as protein concentrates. The protein is low in the amino acids cystine and methionine (Table 4.17). The low and often variable lysine content may be caused by high temperatures during oil extraction in the expeller process (Section 5.14). At elevated temperatures lysine complexes with carbohydrate and is made

Table 4.17. Amino acid composition of oilseed protein (Harvey, 1970). (Reproduced by permission of the Commonwealth Bureau of Nutrition). Figures expressed in mg/g crude protein.

Amino acid	Soybean cake	Cottonseed cake	Groundnut meal (expeller)
Histidine	33	22	22
Isoleucine	50	46	40
Leucine	81	15	64
Lysine	65	34	35
Methionine + cystine	35	30	32
Phenylalanine + tyrosine	86	63	84
Threonine	37	33	29
Tryptophan	15	8.0	10
Valine	51	28	45

unavailable to the animal's digestive system. When the amino acid composition of oilseed meals is compared with the amino acid requirements of animals (Table 4.8), it can be seen that soybean protein almost meets the requirements. It is certainly regarded as one of the best sources of protein for feeding to animals, with methionine probably the major limiting amino acid. Cottonseed and groundnut proteins are not of such high quality, having a markedly lower lysine content as well as lower proportions of the other essential amino acids in the protein.

Probably the best known oilseed is soybean. The flour is a valuable source of protein, the white grades being prepared from dehusked and debittered beans. It can also be used as an extender for cereal flours which are protein deficient. From such flours are made the textured soya protein products such as KESP which are sold as meat substitutes. Indeed, the soybean flours are unusual in that they can be utilized directly by humans. On the whole the other oilseed by-products are fed in quantity only to farm animals. This is partly because some of these, particularly ruminants, may be better able to deal with the toxic elements present, but also because in their present form they are not particularly palatable to humans. However, groundnut flour, specially prepared after oil extraction, can be used for direct human consumption.

4.3.3 Vitamins

When compared with the nutritional requirements of animals, most of the water-soluble vitamins are present in adequate amounts (Table 4.18). The oilseeds are, however, notably deficient in carotene and vitamin E.

4.3.4 Toxic and other substances present in oilseeds

The oilseeds and the protein concentrates derived from them contain a wide range of substances, which, if not destroyed or removed, will have an adverse effect on the nutritional value.

Table 4.18. Vitamin content of oilseed meals (McDonald *et al.* 1973). (Reproduced by permission of the Longman Group Ltd.) Figures expressed in mg/kg fresh weight. Dry matter content: soybean meal, 890 g/kg; groundnut meal, 920 g/kg.

Vitamin	Soybean meal (extracted)	Groundnut meal (decorticated, extracted)
Carotene	0.2	Low
Vitamin E	3.0	3.0
Thiamine	6.6	7.3
Riboflavin	3.3	11.0
Nicotinic acid	27.0	170
Pantothenic acid	15.0	53
Pyridoxine	8.0	10.0

4.3.4.1 Trypsin inhibitors

The best known of these substances are probably the trypsin inhibitors, which are proteins found mainly in uncooked soybeans and other leguminous seeds. The toxic effect results in impaired protein digestion caused by inhibition of the protein-digesting enzymes trypsin and chymotrypsin. These enzymes contain relatively large amounts of the sulphur amino acids, including methionine. Thus, since methionine is the limiting amino acid of soybean, the effect of the inhibitor is compounded by the loss of endogenous essential amino acids already in short supply.

While most of the inhibitor activity can be removed by heat treatment, this in turn can lower protein quality by reducing the availability of lysine and perhaps arginine.

4.3.4.2 Aflatoxins

If groundnuts become infected with certain strains of the mould *Aspergillus flavus*, a group of highly toxic substances known as aflatoxins can be produced. Young animals are most susceptible and poultry rather more so than pigs or sheep. The toxin is carcinogenic with a specific effect on the liver (Section 3.3).

4.3.4.3 Gossypol

The oil-soluble polyphenol gossypol and a number of related pigments are localized in many tiny glands consisting of sacs which are found throughout the kernel of most cottonseeds. Gossypol toxicity is greatest in monogastric animals and, for example, must not exceed 0.016% of the diet of young chicks. Ruminants, on the other hand, are unaffected by quite large amounts of cottonseed meal. The gossypol content is reduced by cooking.

4.3.4.4 Phytohaemagglutinins

The phytohaemagglutinins are a group of proteins found in most legume seeds which, when ingested, reach the bloodstream, where they then combine with glycoproteins of the red blood cell membranes and cause them to agglutinate or coagulate. That from castor beans is called ricin and is extremely toxic. The phytohaemagglutinin from uncooked soybeans has been shown to act as a growth inhibitor in rats (Table 4.19).

Table 4.19. Contribution of the soybean haemagglutinin (SBH) to the growth inhibition induced in rats by diets containing soybean protein (Liener, 1953) (Reproduced by permission of John Wiley and Sons Ltd)

Source of protein in diet	Weight gain (g per 2 weeks)	Growth inhibition (%)
Heated soybean meal	60.0	0
Unheated soybean meal	28.0	43
Heated soybean meal + 0.8% SBH	45.0	26

4.3.4.5 Goitrogens

Rapeseed, *Crambe* and mustard seed contain glycosides which, on enzymatic hydrolysis, release products that are both goitrogenic and growth inhibitory. For example, the enzyme myrosinase, which is present in untreated rapeseed meal, can hydrolyse the sulphur-containing glycoside 3-butenyl glucosinolate to a toxic product, 3-butenyl isothiocyanate:

$$CH_2 = CH - CH_2 - CH_2 - C \begin{matrix} S - C_6H_{11}O_5 \\ \\ N - O - SO_3^- \end{matrix} \qquad \text{3-Butenyl glucosinolate}$$

$$\downarrow \text{myrosinase}$$

$$NSO_4^- + C_6H_{12}O_6 + CH_2 = CH - CH_2 - CH_2 - NCS$$

3-Butenyl isothiocyanate

Even though the hydrolytic enzymes can be inactivated by heating, the glycosides are still present and may subsequently be converted to the toxic goitrogens by intestinal microflora. Obviously, such substances must be removed before these seeds can be passed as suitable for animal consumption.

4.3.4.6 Other toxic factors

Many other toxic substances are found in oilseeds. These include the cyanogenetic glycosides, from which HCN may be released by enzymatic hydrolysis. The amounts present are generally small and in any case normal processing destroys both the enzyme and its glycosidic substrates.

Saponins, which are foam-generating glycosides, are present in soybeans and groundnuts. However, they do not appear to be toxic and, in fact, there is no evidence to suggest that they are absorbed into the body following ingestion.

A number of substances having estrogenic activity are present in soybean. While there is no doubt that one of these, genistein, when fed at levels as high as 0.5% of the diet inhibits growth of rats, it is thought that the levels in soybeans are sufficiently low for any toxic effects to be unlikely.

Phytic acid (Section 4.2.1) is also present in oilseeds and its effect is the same as that described for cereals. It is considered that the increased requirement for manganese, zinc, and calcium in animals fed soybean protein is probably due to metal ion binding by phytic acid.

The presence of many toxic factors in oilseeds and their by-products probably explains why they are less commonly used than they might be for feeding to farm animals and man. Oddly, the widest range of toxic substances seems to be present in soybeans—probably the most popular of the oilseeds for animal feeding. This, no doubt, is a direct result of the time and effort put into the analysis of soybeans and their by-products. In conclusion, the oilseeds and their by-products are potentially a more balanced and complete food than the cereals. Their main disadvantage is the presence of a range of toxic substances which must be removed or inactivated before feeding to animals or man.

4.4 NUTRITIVE VALUE OF GRAIN LEGUMES

Although the *Leguminosae* include approximately 600 genera, with about 13 000 species, only about 10 or 12 are of economic importance today if we exclude soybeans and groundnuts.

The biological value of some species and varieties of legume grains are shown in Table 4.20. These grains contain almost twice as much protein as cereal grains. They also contain a large amount of carbohydrate, most of which is starch (Table 4.21). On the whole, the legume grains are rich in lysine but poor in the sulphur-containing amino acids. Since the reverse is true for cereals, a combination of the two can approach the nutritive value of animal proteins.

Table 4.20. Common names and biological values of some legume grains (Milner, 1975). (Reproduced by permission of John Wiley and Sons Ltd.)

Legume grain	Common name	Biological value
Lens esculenta	Lentil	32–58
Phaseolus aureus	Mung or green gram beans	39–66
Cajanus cajan	Pigeon pea	46–74
Pisum sativum	Pea	48–49
Cicer arietinum	Chick pea, Bengal gram	52–78
Phaseolus mungo	Black gram beans	60–64
Phaseolus vulgaris	Haricot, French or navy beans	62–68

Table 4.21. Nutrient composition of grain legumes before and after cooking (McCance and Widdowson, 1978). (Reproduced by permission of the Controller of H.M. Stationery Office) Nutrient contents expressed in amount per kg of food.

Nutrient	Lentils		Beans (green gram)		Chick pea (Bengal gram)	
	Raw	Cooked	Raw	Cooked	Raw	Cooked
Energy (MJ/kg)	12.9	4.2	9.8	4.5	13.6	6.1
Major constituents (g)						
Moisture	122	721	120	725	99	658
Carbohydrate	532	170	356	114	500	220
Fat	10	5	10	42	57	33
Crude protein	238	76	220	64	202	80
Starch	508	162	344	106	400	168
Minerals and vitamins (mg)						
Calcium	390	130	1 000	340	1400	640
Iron	76	24	80	26	64	31
Carotene	0.60	0.20	0.24	0.44	1.90	2.1
Thiamine	5	1.1	4.5	0.9	0.5	1.4
Riboflavin	2	0.4	2	0.4	1.5	0.5
Nicotinic acid	20	4	20	4	15	5

In general, legume grains are good sources of the B vitamins, but they have no vitamin C activity. Interestingly, sprouted beans, which are a popular food in many countries, have a relatively high vitamin C content.

With the exception of soybeans and groundnuts, the grain legumes can be eaten after cooking directly by man and are a major source of protein and other nutrients in the diets of many people throughout the world. Cooking, which generally requires 3–4 h boiling, after soaking and washing, destroys many of the toxic substances known to be present in the raw seeds.

REFERENCES

General references

Davidson, S., Passmore, R., Brock, J. F., and Truswell, A. S. (1975). *Human Nutrition and Dietetics*, 6th ed., Churchill Livingstone, Edinburgh.
Haresign, W. and Lewis, D. (1978). *Recent Advances in Animal Nutrition—1978*, Butterworths, London.

Specific references

ARC (1975). *The Nutrient Requirements of Farm Livestock, No. 1, Poultry*, HMSO, London.
DHSS (1969). *Report on Public Health and Medical Subjects, No. 120*, HMSO, London.
Eggum, B. O. (1973). *Biological Availability of Amino Acid Constituents in Grain Protein. Proceedings of a Research Co-ordination Meeting 'Nuclear Techniques for Seed Protein Improvement'*, IAEA, Vienna.

FAO/WHO (1973). *Energy and Protein Requirements*, FAO Nutrition Meetings Report Series, No. 52, World Health Organization, Geneva.

Harvey, D. (1970). *Tables of the Amino Acids in Food and Feedstuffs*, 2nd ed., Commonwealth Agricultural Bureaux, Farnham Royal.

Liener, I. E. (1953). Soyin, a toxic protein from the soybean. *J. Nutr.*, **49**, 527–539.

MAFF (1975). *Tables of Feed Composition and Energy Allowances for Ruminants*, Ministry of Agriculture, Fisheries and Food, Pinner, Middlesex.

McCance, R. A. and Widdowson, E. M. (1978). In A. A. Paul and D. A. T. Southgate (Eds), *The Composition of Foods*, HMSO, London.

McDonald, P., Edwards, R. A. and Greenhalgh, J. F. D. (1973). *Animal Nutrition*, 2nd ed., Oliver and Boyd, Edinburgh.

Milner, M. (1975). *Nutritional Improvement of Food Legumes by Breeding*, Wiley, New York.

NAS (1969). *United States – Canadian Tables of Feed Composition*, National Academy of Sciences, Washington, D.C.

NRC (1974). *Recommended Dietary Allowances*, 8th revised ed., National Academy of Sciences, Washington, D.C.

Pirie, N. W. (1975). *Food Protein Sources*, Cambridge University Press, Cambridge.

Sandstedt, R. M. (1965). Fifty years of progress in starch chemistry. *Cereal Sci. Today*, **10**, 305–315.

Scottish Agricultural Colleges (1978). *Nutrient Allowances for Cattle and Sheep*, SAC Publication No. 29, Scottish Agricultural Colleges, Edinburgh.

Whittemore, C. T. and Elsley, F. W. H. (1977). *Practical Pig Nutrition*, Farming Press, Ipswich.

Widdowson, E. M. and McCance, R. A. (1954). *Spec. Rep. Ser. Med. Res. Council*, No. 287, Medical Research Council, London.

Chapter 5

Seed processing

5.1 INTRODUCTION

Although a wide range of products are made from seeds, it is possible to distinguish between two main types: the complex product in which the bulk of the seed is involved (Sections 5.2–5.12) and the fractionation product where a specific component of the seed is extracted and used in the processing (Sections 5.13–5.15). Complex products tend to have complicated chemical and physical properties and include most of the traditional foods made from seeds. The relationship between seed quality and product quality is often ill-defined. Fractionation products, whilst including several traditional food items, also include some newer uses and commonly the relationship between the chemical composition of the fraction extracted and the properties of the product is well understood.

5.2 MALTING AND THE PRODUCTION OF ALCOHOLIC DRINKS

Any cereal grain can be malted but the fermentation industries rely heavily on malted barley, usually referred to simply as malt. For a variety of reasons barley (*Hordeum vulgare*) seems to be the most suitable cereal for this purpose and has become widely cultivated around the world. Commonly, husked barleys are found in Europe, North America, and Australia and include the two-rowed (*distichum*) or six-rowed (*hexastichum*) types. In many parts of Asia huskless varieties, or 'naked' barleys, are found. The husk is formed by the fusion of the palea and lemma in the developing grain. In naked barleys this fusion does not occur and the 'husk' tissues are lost during threshing as in wheat and rye. Whilst of no apparent significance in the malting of mature grain, the husk may be an important barrier to uptake of O_2 in freshly harvested grain (Section 3.6) and pieces of husk help in the filtration of wort from the mash tun. Furthermore, the husk appears to be the more or less sole source of phenolic compounds in malt and, whilst too high a level of this type of material is undesirable in beer as it is a major contributor to non-biological haze formation, too low a level results in an insipid beer lacking in palate and mouthfeel.

Within the husk is the fused pericarp-testa layer covering the starchy endosperm with its surrounding aleurone cells, and the quiescent embryo (Fig. 5.1). The embryo and aleurone layer are important in the formation of enzymes during malting whilst the endosperm, which forms the bulk of the grain, is the main depository of potentially fermentable material. The endosperm is a dead tissue and consists of a tightly packed mass of cells. Starch granules, either large (25–30 μm diameter) or small (5 μm diameter) fill the cells and the spaces around the granules are filled by an amorphous protein matrix.

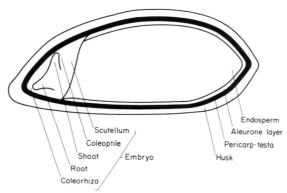

Figure 5.1   Diagram of a longitudinal section through a mature barley corn. The barley corn consists mainly of starchy endosperm with the small embryonic axis attached at one end via the scutellum. The endosperm is completely surrounded by the aleurone layer, a metabolically active zone one to four cells deep. Surrounding both embryo and endosperm is the fused pericarp–testa, which provides a major permeability barrier to water and solutes. Husk material encloses the entire seed. The micropyle (not shown) is a tiny hole in the husk and pericarp–testa in the vicinity of the coleorhiza

5.2.1 Storage of barley

Because of the way in which the outer layers of barley are formed during maturation, barley seeds always contain a certain amount of fungi, even if they have been surface sterilized. These fungi are probably the main cause of deterioration under practical conditions but can be successfully held at bay for several years so long as the moisture content of the seed is sufficiently low. Low temperature is also a factor in storage and the equation given in Section 3.2, connecting temperature and moisture content to longevity, was originally devised for barley. In the case of malting barley, of course, not only must fungal deterioration be avoided but also a very high germinative capacity must be maintained, since an essential aspect of successful malting is even, complete germination of the corn. With modern farm drying equipment there is no theoretical difficulty in reducing barley moisture to

11–12%, which is sufficient for storage without causing any loss in viability. Damage to the embryo can arise, however, where high drying temperatures are used. A frequent cause of later spoilage is uneven drying which leaves a moist spot in the silo, where fungi can grow and start off the whole deterioration process. A second type of problem specific to the storage of malting barley is in the handling of pre-germinated barley. In cold, wet summers many barley varieties begin to germinate on the ear. If this process has gone too far before harvest the subsequent drying will kill the embryos, making the batch useless for malting. Many intermediate levels of viability can be produced in years when pre-germination is common but any fall below 96% viability is likely to be regarded as unsatisfactory by the maltster.

5.2.2 Biochemistry of barley germination

A mature barley seed will remain quiescent for many years at low temperatures and low moisture content. If, however, water becomes available within a suitable temperature range (5–35 °C) the water is taken up through the husk and micropyle and both the embryo and endosperm become hydrated. The colloidal swelling of the endosperm protein, which is not dependent on the seed being alive, results in a distinct increase in size. This puts an internal stress on the integrity of the covering layers. Availability of water, and with it oxygen, to the embryo results not only in a chemico-physical size increase but also a turning on of active metabolism. The seed changes from a quiescent structure with barely measurable gas exchange to an actively metabolizing organism. Although all the details of the initiation of metabolism are not available, it seems that the first increases in metabolic rate are concerned with the amphibolic pathways of glycolysis, the pentose phosphate pathway, and the tricarboxylic acid cycle. Behind this follows biosynthesis of nucleic acids and eventually protein. Afterwards, increase in size of the embryo can occur by cell division rather than just by water uptake and elongation. Some evidence suggests that it is about this stage where cell division is beginning that the developing embryo becomes sensitive to heat damage and desiccation. The first phases of embryo growth are supported entirely on the embryo's own reserves of carbohydrates (mainly sucrose), lipids, and proteins. One of the distinctive features of this early phase, however, is the synthesis of the hormone gibberellic acid (GA), which passes via the scutellum into the cells of the aleurone layer, which it stimulates to produce a wide range of hydrolytic enzymes that are secreted into the endosperm. The enzymes known to be synthesized in response to GA are shown in Table 5.1. Endo-β-1,3-glucanase, arabinosidase, β-amylase, α-glucosidase, phytase, carboxypeptidase, and aminopeptidase also become active in the endosperm but are not under the control of plant hormones. Of this group, β-amylase and carboxypeptidase are always present in the endosperm and merely become activated during germination, whilst the other enzymes are secreted by the aleurone cells.

Table 5.1. Enzymes whose synthesis is stimulated by gibberellic acid during the germination of barley. All of the enzymes listed are synthesized in the aleurone layer and secreted into the endosperm.

Endo-barley-β-glucanase	Endoxylanase
Cellobiase	α-Amylase
Laminaribiase	Limit dextrinase
Exoxylanase	Endopeptidase

The combined action of these and possible other enzymes, results in what the maltster calls 'modification' of the endosperm. This process shows itself most easily in the conversion of the hard, brittle barley corn into the friable, easily crushed malt corn with a loss of about 10% in dry weight. Enzymic action appears to be sequential, in that proteolytic attack seems necessary before amylolysis can begin, and the small starch granules become digested almost entirely before any significant attack on the large starch granules has occurred. Sections through malt corns show that the endosperm is no longer a densely packed tissue but that much of the cell wall material and matrix protein, as well as the small starch granules, of the barley endosperm have gone.

5.2.3 Dormancy in barley

Normally, mature barley grains will not germinate until after a few weeks or months of after-ripening. This primary dormancy can be of great significance to the maltster, as the move on to the new barley crop occurs in the early winter. Successful malting requires a rapid and even germination of over 95% of the corn and is impossible with samples which show a significant degree of dormancy. The typical test for barley germination is to place 100 corns in a petri dish containing two filter-papers and 4 ml of water. The number of seeds sprouted after 3 days gives the percentage germination (germinative energy) of the batch of seeds. Fully after-ripened seeds give values of over 96% whereas dormant seeds may only give very low figures in this test. However, a forcing test employing 0.75% hydrogen peroxide normally induces the full germinative response of dormant seeds. Basically with dormant seeds the maltster has to wait until dormancy recedes before attempting full-scale malting, although variation in the steeping technique, particularly the use of air rests where the water is temporarily drained away from the grain, can produce a high level of germination where otherwise only a relatively poor response would be achieved. It has been reported that dormancy can be destroyed on a commercial scale by drying the barley down to near 10% moisture. An alternative procedure involves holding the grain at 40 °C for several days followed by slow cooling back to ambient temperature. These techniques appear to succeed even in particularly bad years when the newly harvested barley fails to respond to the introduction of air rests into the steeping programme. Excess water on the seeds can reduce germination in batches which are apparently fully viable according to the 4 ml test. In order to detect this the percentage viability in a test using 8 ml of water is often reported. This reveals the so-called water sensitivity, a phenomenon which, like normal dormancy, passes after an after-ripening period.

The causes of dormancy in barley are not known for certain but the evidence supports the idea that the husk and seed coat in some way impede the passage of water and oxygen to the embryo. Thus, hydrogen peroxide can stimulate dormant seeds into germination, as can physical removal of the husk from around the embryo. Many chemicals which break dormancy are metabolic inhibitors whose action in slowing down respiration could lead to a build-up of oxygen within the seed, so allowing some critical oxidative step to occur. However, much of the evidence could be interpreted in other ways, e.g. the treatments mentioned could cause leaching out or oxidation of a natural germination inhibitor whose concentration would normally fall for other reasons concerned with the metabolism of the seed. A further complication arises from the observation that occasionally when dormancy levels in barley are very high even complete removal of the husk fails to achieve full germination, although sufficient after-ripening does. This implies that there is more than one cause of dormancy and it seems plausible that during after-ripening this changes successively from aspects of the intrinsic chemistry of the embryo to an external restriction on gas exchange imposed by the husk.

5.2.4 The malting process (Fig. 5.2)

In the traditional malting process barley is soaked in a trough with water until fully imbibed. The water is then drained away and the wet barley spread evenly over the malting floor to a depth of 4 inches. This mass of grain is allowed to germinate with conditions maintained constant as far as possible, although in practice the only controls are likely to be the opening and closing of windows to control the temperature together with the manual turning of the barley to maintain an even moisture content. After about 8 days of growth the acrospire is likely to be about to break out from the seed. This process is known as chitting, and the malt is then considered to be ready. The process is completed by transfer of the green malt to a kiln, where it is dried down to a very low moisture content (3–5 %) to stabilize the product. Kilning is a very important part of the process as the heating greatly reduces the activity of the hydrolytic enzymes and causes development of colour and flavouring compounds due to interaction between free sugars and amino acids.

Modern departures from the traditional process have not altered the principles of malting but rather have taken advantage of machinery to turn the malt and thereby use deeper beds with all the consequent savings in space and labour. In mechanically turned systems such as Saladin boxes, the depth of the grain may be 3–4 ft as opposed to the 4 inches of the malting floor. The latest type of malting box is equipped with automatic sprinklers to maintain moisture contents and fans which can blow attemporated air through the whole bed of grain for cooling. This also allows kilning without movement of the grain. These techniques have led to a reduction in the time necessary for malting to 5–6 days. A further reduction in time can be achieved if a maltster is prepared to add low concentrations of gibberellic acid. Gibberellic acid is the natural hormone of barley which stimulates the aleurone layer to produce many of the hydrolytic enzymes, and its addition can accelerate the process by a whole day. More substantial gains are achieved, however, if the barley is scarified, or abraded, before malting. In this process an abrading machine is used to remove a small portion of the husk from the distal end of the grain so allowing the gibberellic acid direct access to the end furthest from the embryo. This means that modification proceeds from both ends of the grain simultaneously and so the required degree of modification can be achieved in a much reduced time, e.g. 3 days.

Barley appears to be the only grain used today for malting on a large industrial scale, but there is currently great interest in Africa in extending the potential of sorghum malt. Sorghum grows widely in the warmer parts of the world and is used in the USA as an unmalted adjunct by brewers, but the traditional African beer is based on malted sorghum. As normally practised, the malting of sorghum is a fairly crude process in which the grain is steeped and allowed to sprout at ambient temperatures for about 2 days. The green sorghum malt is then crushed and used directly in the two-stage fermentation process. There is much interest in applying more controlled conditions of malting to optimize yield and also to introduce kilning so that a stable product can be made. It seems likely that a great deal of biochemical and physiological information on sorghum germination and malting will soon become available.

5.2.5 Mashing (Fig. 5.2)

Mashing is the first stage of the brewing process, in which the breakdown of the endosperm reserve material, begun in malting, is completed. Some knowledge of this process and the subsequent stages of brewing and distillation is necessary before the desirable properties of a good malt can be appreciated.

The exact mashing procedure used depends on the type of malt and the eventual product of the process. Well modified malts, popular in the British Isles, are mashed in the infusion system which uses a single temperature stand, usually 65 °C, to complete modification and solubilization. The milled malt is mixed with hot water and then allowed to stand in a mash tun for the time necessary to remove all the starch from the grain. The major enzymic activity occurring is α-amylolysis. α-Amylase, which breaks the bonds of α-1,4-linked glucans at random, is a relatively heat-stable enzyme and under the partially protected conditions of a thick infusion mash is able to work for several hours, producing a wide range of sugars from glucose upwards,

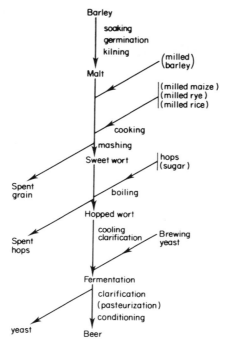

Figure 5.2 Flow diagram for malting and brewing. Items shown in parenthesis are not obligatory parts of the process

including higher dextrins whose α-1,6-links have prevented further attack by the enzyme. β-Amylase, which removes maltose units from the non-reducing end of α-1,4-linked glucans, is also active during mashing but as it is not as heat-stable as α-amylase it does not normally survive the whole process. Nevertheless, as opposed to the complex enzymic situation during malting, the main reaction occurring during infusion mashing is the combined action of α- and β-amylases on starch and smaller polysaccharides. Once mashing is complete an important difference between breweries and distilleries occurs. The brewer, largely for historical reasons, is interested in maintaining a complex carbohydrate profile in the wort so that a distinct amount of material is not fermentable and will pass unchanged into the beer where it is thought by many, if not all, to impart desirable qualities of fullness. Thus, a brewery wort is boiled and this denatures all remaining malt enzymes, so saving the more complex carbohydrates from further degradation. The distiller, on the other hand, is not interested in the organoleptic properties of the distillery wash but rather in the need to convert as much of the cereal starch as possible into alcohol. This means that the highest possible yield of fermentable sugars must be achieved. Thus, the extract from the mash tun is not boiled, but yeast is added directly so that, even during fermentation, the malt amylases and debranching enzymes are still active in reducing the molecular size of the polysaccharides.

On the continent of Europe, lager brewers have traditionally used under-modified malts and they have developed the decoction mash system to deal with this type of material. Under-modified malt can be produced with less malting loss due to respiration and seedling growth than is the case for well modified malts. However, such malt is generally deficient in several growth factors required by the yeast. Most notable of these are the amino acids and, as malt proteolytic enzymes are rapidly destroyed at 65 °C, little liberation of α-amino acids can occur in infusion mashing using undermodified malt. The problem is overcome in decoction mashing by starting at 45–50 °C, the so-called protein rest, which allows proteolysis to occur. The temperature is then raised to about 65 °C to complete amylolysis. The rise in temperature is achieved by removing part of the mash, boiling it, and returning it to the bulk of the mash. As long as the relative volumes are carefully controlled, a tight regulation of temperature can be achieved.

The previous paragraphs have considered mashing as a process using only malt. Whilst this may sometimes be true, in many cases unmalted cereals are included in the mash tun together with the malt. This is possible because commercial malts normally have more enzymic capacity than is necessary to break down their own reserve material and so can convert additional starch to fermentable sugars. The extent to which these cereal adjuncts can be used is limited by the enzymic capacity of the malt, particularly the α-amylase, unless the brewer chooses to add microbial enzymes to the mash. As long as no significant technical problems present themselves and the product is satisfactory there is economic sense in supplanting malted by unmalted grain as the latter is likely to be cheaper. In many countries the exact extent to which malt is used is determined more by legislative definition of beer and spirits than by technical considerations.

5.2.6 Criteria of grain quality

The major considerations in the purchase of grain are always the cleanliness and wholesomeness of any batch, together with its cost. The latter point should be calculated as the cost of the eventual product derived rather than the actual price of the grain to take into account variations in grain composition and quality.

(a) Grain for malting

Barley for malting should be a variety which is generally recognized for good malting quality. The grain should have a percentage germination in the standard Institute of Brewing 3-day 4 ml test of 96% or over and, furthermore, the corns should be reasonably large and of a uniform size. This greatly aids in the evenness of modification during malting. Another common requirement is a restriction on the nitrogen content as determined by the Kjeldahl method. Brewers tend to prefer lower nitrogen barleys for two main reasons: firstly, the lower the nitrogen the higher is the starch content and thus the possible extract, and secondly, high nitrogen contents are associated with the problems of non-biological haze development much later in the packaged beer. The use of non-malted cereals in mashing does tend to reduce the significance of malt nitrogen, however, as these adjuncts contribute sugars to the wort, but relatively little by way of nitrogenous compounds.

(b) Malt

Some malt qualities, such as colour, are arbitrarily set by the type of beer to be produced, whilst other properties are more universal. Foremost amongst the latter is that extraction of starch during mashing should be complete, as the resultant sugars are the main contributors to the specific gravity of the wort. In the brewery this will be measured in litre degrees per kilogram of grain. This measurement has replaced the earlier unit of pounds of extract per quarter of grain. The cost of a litre degree is the key figure in comparing the real costs of various ingredients of the mash tun. Also very important is the percentage fermentability of the malt extract; this should be appropriate to the brewer's standards and is commonly around 75%. Although it is possible that a wort could cause defective or faulty fermentation owing to lack of a specific component, e.g. zinc or fatty acids, the only biochemical criteria commonly applied are that the total nitrogen content should not be too high and that the permanently soluble nitrogen, mainly amino acids, should be sufficiently high to give the desired yeast growth but not so high that substantial amounts of amino acids persist into the beer. A high amino acid content makes beer more liable to bacterial infection, particularly by *Lactobacilli*. This latter point is, of course, much more significant for naturally conditioned beers than for those which are pasteurized.

(c) Unmalted cereals

Almost any source of starch can be used as a mash tun adjunct but, because beer and whisky are traditionally cereal-derived products, cereal adjuncts are preferred today. Such adjuncts act as sources of fermentable sugars rather than yeast nutrients, although the amount of nitrogen obtained does depend on the cereal, e.g. maize gives rise to very little nitrogen whilst raw barley can contribute distinct amounts particularly if a decoction or temperature programmed mash is used. A technical point here, which may be decisive for a brewer, is that, as the gelatinization temperature of all cereal starches other than that of barley is higher than 65 °C, most unmalted cereal adjuncts other than barley have to be cooked to break open the starch particles before the grain can be utilized in the mash tun. In the case of maize flakes the starch granules are disrupted during the flaking process and so they can be used directly.

The main cereals used in unmalted form are maize (*Zea mays*) and rye (*Secale cereale* L.). Rye is normally included together with barley malt in the mash for gin and various forms of schnapps popular all over Northern Europe. Grists containing up to 80% maize are used in the manufacture of grain whisky in Scotland whilst American Bourbon whiskey must include at least 51% maize. A typical recipe for Bourbon would be 60% maize, 28% rye and 12% barley malt. As a group, distillers tend to be more bound by traditional recipes than brewers, partly at least because of the greater difficulty in experimentation with the flavour of spirits but also because the definition of spirits is much more tightly defined in terms of the original grists.

When considering individual batches of cereal, apart from the usual criterion of wholesomeness the most important point seems to be that, as for malting barley, the grain should be of sufficient size and also uniform in size distribution. This makes for good milling performance, which, particularly in the case of raw barley which is used uncooked, is the major criterion in ensuring a good extract in the mash tun.

Alcohol for industrial purposes can be made by fermentation of acid-hydrolysed starches normally obtained from maize. Recently, however, hydrolysis of the starch by suitable bacterial and fungal enzyme, themselves produced on an industrial scale, has become a relatively attractive process. It seems likely that the importance of this conversion will grow as the sources of cheap raw materials for the chemical industry decline.

5.3 MALTING AND THE PRODUCTION OF NON-ALCOHOLIC DRINKS

Although malting is traditionally the first step in the production of a variety of alcoholic drinks as outlined above, in more modern times other uses have been found for malt. The sweet wort from a mash can be concentrated to a thick syrup and sold directly or, with additives such as vitamin A, as a tonic or health food. In some countries the sweet wort is carbonated and bottled and sold as a drink (malzbier). This may be in direct competition with traditional beer and to many people it is an acceptable alternative to beer, particularly in countries where there is a high awareness of the potential physical and legal consequences of driving with a high blood alcohol level or where there is a religious objection to alcohol. Malt and malt extract are used as flavouring agents in some types of loaves, biscuits, and breakfast cereals. Several proprietary powders sold for making into drinks, particularly bedtime drinks, by mixing with hot milk are largely malt and cereal-based products involving an extraction phase very similar to that used in brewing.

By and large, the analytical requirements for malt destined for use in the manufacture of non-alcoholic products are not so stringent or restricted as with malt intended for the brewing industry, although in some cases colour may be very critical. This is largely because in the non-alcoholic products a well developed malty taste is the main requirement and this is not difficult to achieve. The brewer, however, is not only concerned with the consistent extraction of fermentable sugars and yeast nutrients but also in achieving reproducible fermentation patterns so that a reliable beer flavour is obtained. Furthermore, the length of time that a beer remains clear, and thus its shelf-life, depends on the original malt quality as well as various aspects of the processing and packaging.

5.4 NON-BEVERAGE PRODUCTS FROM MALT

As well as the main product, malt, the malting process results in the formation of a quantity of rootlets which are separated from the grain after kilning. These dried rootlets have a relatively high protein and fibre content and, often together with malt dust, can be utilized in the manufacture of feed for ruminants and horses.

Similarly, although the main product of the mashing process is sweet wort, the residual grain (spent grain) also has a high protein and fibre content and is a valuable ingredient in cattle feed. In many cases the spent grain can have the excess yeast added to it before drying and this can increase the nutritive value still further although, of course, nucleic acid levels must not be allowed to become too high. Because in distillation only the coarsest parts of the grain are removed before fermentation, the residue after removal of the alcohol always contains a mixture of grain and yeast materials.

Recent reports have indicated that brewers' spent grain can be incorporated at up to 15% in certain biscuits for human consumption without an adverse effect on flavour, and thereby boost the protein and fibre content of the product.

5.5 SAKE

Sake is a traditional alcoholic drink of Japan and is produced from rice. In terms of alcohol production sake is equivalent to beer in Japan although, because of its much higher alcohol concentration (ca. 16%), the volume of sake produced is much lower than that of beer. It is claimed that only rice grown in Japan gives good-quality sake, although some Korean, Taiwanese, and Californian short-grained rices can be used. Little seems to be known at a chemical and biochemical level about what makes a good sake rice, but there are many practical indicators of rice quality. The grain should not have too high or too low a moisture content and levels between 12 and 17% are preferred. Grain which has been overdried is unlikely to be satisfactory for sake manufacture, even if moisture is re-absorbed. The main constituents of sake rice are starch (72–73%), protein (7–9%), fat (1.3–2.0%), and ash (1.0–1.5%). Of more immediate use in assessment of the grain are a number of features depending mainly on the size and appearance of the grain. The brown rice should have a good lustre and not include discoloured kernels, although a few purple kernels do no harm. Large, full grains are the best but small-grained rice which is of good proportions can be used. Particularly important here is that the groove in the kernel should not be too deep (50–50 μm) and the bran layer should not be too thick (42–54 μm). On inspection of a sectioned grain, preferred rice should show a white haze in the centre rather than the outside. This effect is characteristic of large grains but does occur to a certain extent in smaller grains.

5.5.1 Preparation of the rice (Fig. 5.3)

Rice has to be carefully prepared before it can be used in any of the stages of sake brewing. The first job is polishing. This is done in a vertical mill using a roller composed of carborundum and feldspar, and involves removal of a specified percentage of the outer layers of the grain. In comparison to rice for food purposes, where 90–92% of the original weight is retained, only 70–80% is retained for sake brewing and the polishing rate may be as low as 50%

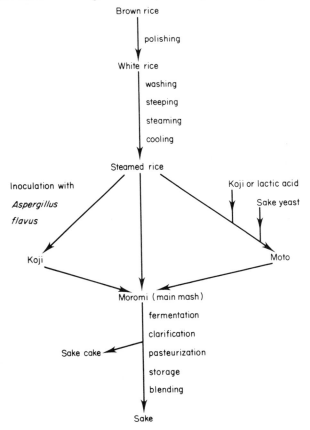

Figure 5.3   Flow diagram for sake manufacture

retention for high-quality sake. It is very important in polishing that the mill is adjusted to take into account the size and hardness of the grain, otherwise irregular polishing or excessive breakage of kernels can occur. The effect of the polishing process on the overall chemical composition of the grain is variable; the percentage of starch increases whilst the fat content is virtually eliminated by removal of the first 20%, which includes the embryo and aleurone. The ash content also shows a very rapid decline with removal of the outer layers, but not as dramatic as the changes in fat content. Percentage protein content tends to decline in proportion to polishing rate. On the physical side, the kernels after polishing are of uniform size and smooth exterior with no trace of the outer brown layers, embryo, or groove. It is the polishing process which sets the demand for large-grained rice of even size as the polishing rate for any given quality of sake has to be increased with small or irregular grains.

Once polished, the rice is washed to remove all traces of bran, and this is followed by steeping in water at 15 °C. The aim of this process is to obtain a 25–28% moisture content, which is considered ideal for the next part of the process. The rate of water absorption can be very variable, depending on the quality and degree of polishing of the rice. Steeping may take 20 h for hard grain whilst 2 h will be sufficient for soft grain. Grain polished up to 50% may take up sufficient water in a few minutes.

When the required water content has been reached the grain is allowed to drain for several hours before being piled into a steaming vessel. Steaming times are usually of the order of 20–60 min after steam first appears through the bed of rice, but the exact time has to be judged carefully depending on the type of rice, the polishing rate, and the steeping conditions. After steaming, the moisture content of the grain is increased by about 10%, and the kernel, whilst still resistant to crushing between the fingers, becomes slightly translucent and sticky. Once the correct degree of steaming has been achieved the rice is normally cooled by a forced air draught.

From a chemical point of view steaming gelatinizes the starch particles and denatures the protein, making both types of polymer much more susceptible to enzymatic hydrolysis. An even water content is very important during steaming as this ensures that the heat is transmitted throughout the grain so there is no possibility of enzymically resistant areas remaining.

5.5.2 Modification of the rice (Fig. 5.3)

The steamed rice contains almost no material fermentable by yeasts, nor has it any enzymic content. Because of this it is necessary to introduce hydrolytic enzymes from another source and this is traditionally done by use of strains of *Aspergillus flavus*. Some of the steamed rice is cooled to 35–40 °C and transferred to a culture room at 25–30 °C and 80–90% relative humidity. Here a proportion of *tane koji* is introduced. *Tane koji* is steamed rice made from kernels with a very low polishing rate (98–99% retention) which has been inoculated with spores of the appropriate strain of *A. flavus*. The fungus spreads fairly rapidly throughout the mass of steamed rice and the temperature is likely to rise to about 40 °C before the *koji* is considered complete. This process takes about 2 days. *Koji* is typically white in colour and relatively dry. Fungal mycelium is present on the surface and throughout the grain.

5.5.3 Fermentation

At the same time as *koji* is being prepared, about 7% of the steamed rice is taken for *moto* preparation. This is essentially a yeast starter fermentation which has either been allowed to acidify by addition of some *koji* or deliberately acidified by addition of lactic acid. Once the bacterial fermentation has run its course in the traditional process and wild yeasts have more or less completely disappeared, a pure culture of sake yeast is introduced and a vigorous alcoholic fermentation ensues. After several days the *moto* is cooled, rested for a few more days, and is then ready for the next stage of the process.

The final stage of sake manufacture is the main mash or *moromi*, in which *koji*, *moto*, and the bulk of steamed rice which has been stored at 10 °C are combined with additional water. The process is usually carried out in a stepwise fashion involving three additions of *koji* and *moto*. Once the third addition is complete a long, slow fermentation lasting nearly 4 weeks sets in. Throughout the process the rice starch is degraded to soluble sugars, which in turn are fermented to alcohol by the sake yeasts. The temperature is usually kept between 15 and 18 °C and the final alcohol concentration is usually in the range 16–20%.

The complete *moromi* mash is filtered through small linen or synthetic fibre bags to yield fresh sake, which must be further processed before being suitable for sale (Fig. 5.3). The residue left in the bags after pressing is referred to as sake cake (*sakekasu*) and is used for human consumption in a variety of ways. Typical uses include *amazake*, which is a hot drink made in the home from *sakekasu*, sugar, and hot water. Another domestic product is *sakeshiru*, a form of soup in which *sakekasu* is boiled with vegetables, meat, and water. Although brewers' spent grains have been used in the West as cattle food they have not been traditionally used for direct human consumption. This may be changing as some biscuits are now available, containing up to 15% of spent grains which are relatively high in protein and fibre content.

In terms of approximate yield, 1000 kg of white rice would be derived from 1100–2000 kg of brown rice, depending on the polishing rate, and would give 2500 litres of sake and 250 kg of sake cake (55% moisture).

5.6 WHEAT PRODUCTS

Wheat is probably the largest crop grown in the world in terms of yield of dry material; the bulk of it is derived from varieties of the single species *Tritium aestivum*. Amongst the handful of other species grown commercially *T. durum* has a special place as the source of paste products. Apart from the species, three factors are of special significance to the farmer. Firstly, the growing habit of the variety is important; seed sown in the spring (spring wheat) tends to harvest late and is likely to be inferior to seed sown in the autumn (winter wheat), where the small plant becomes established before the winter and so makes an early start in the following year. Winter wheat cannot be grown, of course, in areas where there are severe winter frosts and there is the possibility of significant damage by pests. However, in recent years winter wheat has gained in popularity, largely owing to its higher yielding qualities. As well as yield the grain quality is also very important and wheat is normally divided into hard or soft, white or red categories. No sharp divisions exist, however, as agricultural conditions and the weather can be more important determinants than variety. The best wheats for making leavened bread are commonly believed to be the hard red spring wheats from Canada, and much of the international trade is in wheats of this type, which are brought by many countries, particularly those in Western Europe, for mixing with their own native soft wheats. Hard wheat can only be grown under relatively dry conditions and, as the bulk of the wheat crop is made into leavened bread in the West, hard wheats tend to be at a premium on the world market, particularly if there has been a wet year in the major grain-growing areas of North America.

5.6.1 Criteria of grain quality

(a) General tests

The qualities looked for in a batch of wheat are in many ways the same as those for any seeds. It is important that foreign matter and broken seeds are at a minimum; clean, bright, reasonably large, even-sized grain is essential for all food uses. As well as visual inspection, simple criteria are the weight per unit volume (kilogrammes per hectolitre), 1000 corn weight, and the proportion of corns which will not pass through standard wire-mesh sieves. All of these parameters can be used empirically to predict the likely milling yields.

(b) Colour

Wheat is commonly classified as red or white, depending on the colour of the outer layers of the bran. Colour is basically a varietal property although it can be affected by environmental factors. Most of the commercially grown varieties are of the red type, but white wheats are popular in the East where they are used largely for the production of unleavened bread. A small amount of hard white wheat is produced in the USA and Canada for breadmaking and the white durum wheats are preferred over the red for production of macaroni and pasta.

(c) Breadmaking ability

The essential property which separates wheat from all other cereals is the ability of the flour to be used in the manufacture of light, well risen, even-textured leavened bread. Whilst it is possible to test-bake wheat flours to assess truly their quality, other tests are needed in order to handle the large numbers of samples available in the market. Three properties are normally associated with good breadmaking quality: vitreousness of the kernels, hardness of the grain, and protein content. Vitreousness is a very subjective character and can be used only as a rough, though simple, guide. In many cases, hardness is also a subjective character based on the appearance of the corn rather than any physical measurement. However, objective measures of hardness are available. A standard quantity of wheat can be pearled in a laboratory pearler for a standard length of time and the amount of material removed reported as a percentage of the weight of wheat used. This is referred to as the pearling index. The higher the index the softer is the grain. An alternative approach to measuring hardness is to assess the energy consumed whilst grinding the grain in a burr mill. This gives the wheat hardness index.

The protein content of wheat measured by the Kjeldahl technique is a property which is much more reliably determined than vitreousness or hardness and is an essential item in the description of a wheat consignment. To a large extent protein content correlates positively with bread-forming ability and also with suitability for a range of other purposes (Table 5.2).

Table 5.2. Uses of wheat in relation to protein content

Protein content (at 14% moisture)	Classification	Product
8–10	Soft	Biscuits, cakes, pastry
10–11	Soft	Crackers
12–14	Hard	Bread
13	Hard (durum)	Semolina, macaroni, pasta

Whilst the protein content is important in determining the suitability of a wheat for breadmaking, it is not the only criterion. The quality of the protein is also important (Sections 4.1.1 and 4.2.2). Other components of the grain may also be significant here, as a bread dough is a highly complex physico-chemical system. Within the intact cereal endosperm the various components are located in different parts of the cell, thus preventing certain spontaneous chemical reactions. In dough manufacture, which is the first stage of bread production, the grain is ground, aged, water and other materials are added, and the whole mixture is thoroughly kneaded. This process results in formation of a sticky elastic substance known as gluten, which can be isolated from the dough by washing out the starch. However, although gluten is composed of about 80% storage protein, small amounts of other proteins such as albumins and membrane proteins, lipids (5–10%), and carbohydrate (10–15%) are present and the influence of these compounds is not well understood. The possibility also exists that the starch itself may have some influence on the way in which the loaf forms.

Because of the complexity of the dough itself, some assessment of the breadmaking quality or 'strength' of a flour is necessary. A popular technique is a variant of a sedimentation test in which coarsely ground flour, which has had the bran sifted out, is suspended in an aqueous lactic acid solution in a measuring cylinder. After standing for 5 min the volume of the swollen flour which appears as a 'sediment' at the bottom of the cylinder is read to give a sedimentation value. The higher the value the stronger is the flour.

(d) α-amylase content

Mature wheat, like other cereals, contains β-amylase activity but no α-amylase. This enzyme is produced during germination of the seed. However, particularly under wet conditions, wheat is very likely to germinate on the ear (pre-germination) and flour made from batches of grain containing pre-germinated corns will contain α-amylase. Up to a certain point this is considered to be good as some amylase action is necessary in the dough to supply the yeast used in baking with fermentable sugars. In fact, if the native α-amylase activity is too low, bakers normally add malted wheat, malted barley, or microbial α-amylase to ensure the presence of sufficient sugars. However, if the native α-amylase activity is too high an excessive amount of starch can be degraded to α-limit residual dextrins and this results in a flour which will not rise properly and which has a sticky crumb structure.

5.6.2 Storage of wheat

The quality of wheat in store can never be better than that of the freshly harvested grain. Fungal attack in the field, frost, substantial pre-germination, and the harvesting of immature grain all result in a decrease in grain yield and quality, and may lead to problems during storage. As far as storage is concerned all of the factors discussed in Section 3.2 are relevant, but the main property which determines the life of stored grain is the moisture content. This should be below 13.5% for even fairly short-term storage. When long periods of storage are contemplated, such as in the strategic grain stores of several countries, much lower moisture contents are necessary. This is because the moisture content varies locally within a batch of grain; at an average level of 13.5% moisture there may be pockets of grain with moisture levels sufficiently high to allow the slow development of fungi. As the microorganisms develop, their respiration produces water which encourages wider growth, and rapidly deteriorating areas can suddenly develop in the grain after months of storage. Once this has happened it is only a relatively short time before the whole batch becomes infected and useless.

On the other hand, the problem cannot be overcome simply by extensive drying to very low moisture levels. Very dry corns become brittle and are easily broken during handling of the grain. This broken grain is discarded before milling and so represents a loss in yield. Even if dry corn does not break prior to milling, it is likely to shatter in the mill rather than being ground, causing both mechanical problems and a deterioration in flour yield and quality.

The moisture problem is further complicated by the fact that not only is moisture content itself important, but so also is the way in which drying is carried out. It has been known since the 1940s that the germination capacity of wheat deteriorates if certain temperature/moisture relationships are exceeded. The critical temperature can be calculated using the following equation:

$$T = 122 - 5 \log t - 44 \log m$$

where T °C is the critical temperature, t min is the length of time of treatment and m% is the moisture content. The equation implies that the higher the moisture content of the grain then the lower is the critical temperature. Although in most cases viable wheat is not required for industrial utilization, the germinability of stored grain is a good guide to the level of intrinsic damage which has occurred. If anything, loss of baking quality is more sensitive to temperature at moisture contents above 20% than is germinability, and less sensitive in grain below this moisture content.

Checks on the quality of grain in store are commonly done by eye, but in many cases measurement of the fat acidity (Section 3.4) has been found useful. Sound wheat normally gives figures of under 20 mg per 100 g whereas deteriorated wheat is likely to result in values over 100 mg per 100 g. In some cases an increase in fat acidity is accompanied by other indicators of deterioration, but this is not always the case. It seems that fat acidity values alone are not particularly useful data, although they may give a practical guide within a well defined situation.

5.6.3 Milling

Almost without exception the first stage in the utilization of wheat is the grinding of the grain into flour. Thus the milling properties of any batch of grain are crucial to its subsequent usefulness. Before the wheat can be milled, it must be cleaned of small and broken kernels and of objectionable materials such as soil, fungal growths, and a variety of weed seeds. The cleaning can be done without the use of water by various machines which employ sieves, air currents, and beaters to effect the separation, or the cleaning can be effected by washing. Washing is particularly desirable for very dirty, smutty, and dusty grain and is a useful practice if a flour with a low bacterial count is required. The ratio of water to grain is usually between 1.0:0.5 and 1.0:1.0.

During washing the outer layers of the bran rapidly take up water, leading to a 4–5% increase in moisture content. This phase is followed by a slow, steady uptake of water as the liquid penetrates further into the grain. Water pick-up is fastest with small grains of soft wheat and slowest with large grains of hard wheat. In many cases, particularly the hard wheats, this moisture pickup is useful as the optimum moisture contents for milling are higher than for storage. The exact moisture content needed depends on the wheat quality and subsequent processing intended, but levels of 16–20% for hard wheat and 14–15% for soft wheat are common. Although some soft wheats may need to be dried before milling, the hard wheats invariably need addition of water. The optimum moisture content is a compromise between the requirement for moisture to allow some softening of the endosperm and prevent excessive brittleness of the bran on the one hand and a desire for dryness to ease the separation of bran from endosperm and the subsequent sifting operations on the other. As the penetration of water increases with temperature there is an obvious industrial advantage in using warm water. It seems that temperatures up to 46 °C can be safely employed but above this level there is the possibility of damage to the gluten and hence the baking quality of the flour. The relationship between temperature and moisture content cited above (Section 5.6.2) is relevant in hot-conditioning of grain. In some hot-conditioning processes very short bursts of steam are used. If the time is extended and steam is injected for up to 5 min the temperature of the grain can rise to 100 °C and, whilst the gluten is completely denatured, the flour obtained can be used in place of separated starch for a number of purposes, such as soup, confectionery, and adhesive manufacture.

The mills used are almost invariably roller mills of fairly similar design. Differences in the exact pattern of use and in nomenclature occur, depending on the geographical location of the mill and the type of wheat being handled. A full description of milling with all its minor variations is beyond the scope of this book but the basic system is easily understood. The material to be ground is passed between two appropriately spaced revolving rollers. The stream of crushed grain then passes through a series of appropriately sized sieves, allowing various sub-streams to be separated for further milling or collection as a particular grade of flour. A flour mill has many sets of rollers operating essentially in two sequences: the breaking mills and the reduction mills (Fig. 5.4). The main point of the breaking process is to remove bran from endosperm. The rollers have spiral flutes which become finer and finer as the position of the mill in the series increases. At the same time the rollers are set progressively closer together. The rollers in the breaking mills are driven at different speeds. The breaking process removes the bran and produces a series of flours of various particle sizes. These flours are fed to the second series of mills for the reduction process.

The idea is much the same as in the breaking process, except that reduction mill rollers are not fluted and are driven at more or less the same speed. For the best performance the reduction mill must be fed particles of an even size and the result of all this activity is the conversion of most of the wheat into a

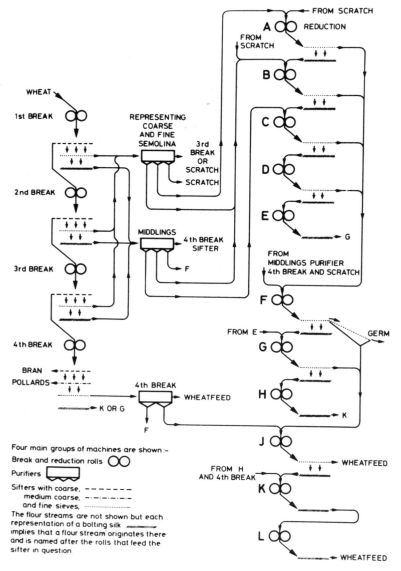

Figure 5.4   Outline of the wheat milling process (Spicer, 1975). (Reproduced by permission of Applied Science Publishers Ltd)

series of flours each with its own characteristic properties. It is the job of the miller to control conditioning of the grain and the operation of the mill to obtain the most desirable results in terms of yield and baking properties of the resultant flour.

Whilst the miller may have room for manoeuvre in determining the yield and properties of the various streams of flour, the overall yield of flour is normally set by the single criterion of colour. Most of the colour of flour is due to the presence of bran. Up to about a 65% extraction rate (the percentage of the original wheat finally collected as flour) there is little effect of extraction rate on colour, but the colour begins to increase disproportionately at extraction rates over 70%. An extraction rate of 92% is likely to yield a flour with twice the colour of that obtained at a 70% extraction rate. Commercial extraction rates for white flour are usually around 72%. The figure is basically determined by the colour requirement, but the relative prices of flour and the material which does not pass the bolting silks, the 'offal', is always taken into account when setting the exact extraction rate.

5.6.4 Flour Products

A wide variety of breads, biscuits, cakes, pastries, and crackers are made from wheat flour and the processes are too numerous to describe in detail here. As has been mentioned in the preceding section, the main criteria in the use of wheat flour for breadmaking are the protein content and the strength of the flour. For most of the other purposes, such as cake and biscuit manufacture, a flour of lower protein content and lower strength is preferred. This is not just a choice by default but is a positive preference for a soft wheat flour for non-bread products; soft wheat flour gives a better texture and spread. Chlorination of soft wheat flour improves cake quality still further. Chlorine reacts with all the major components of flour in a variety of ways but the significant improving interactions are not known for certain. A point which must always be remembered when comparing the two major outlets for wheat flour is that whereas bread is made from a dough which is essentially a flour–water mixture with only minor additions of other materials, cakes, pastries, and biscuits are made from a batter which not only is much more fluid than a dough, but also contains substantial quantities of such materials as milk, sugar, eggs, and fats as well as flour and water. The non-flour components will obviously have major effects on the characteristics and quality of the eventual product, both in their own right and through interaction with the flour.

When it comes to the assessment of a dough or batter, although there are a number of chemical and physical tests which are useful within closely defined situations, the proof of the pudding is still very much in the baking and eating. Test bakes are the only completely reliable method for establishing flour quality for any particular purpose.

5.6.5 Paste Products

The use of wheat in the making of bread and cakes as described in the previous section is essentially a European or European-derived activity. The most widespread technique for the preparation of wheat for consumption is a much simpler process involving a wheat–water paste. In the earliest form of the technique the crushed endosperm is mixed with the minimum of water to form a stiff paste, which is then formed by hand into a convenient shape and cooked immediately. Indian chappatis and Chinese noodles are good examples of this type of paste product. Although any kind of wheat can be used it has been observed that the best results are obtained when semolina rather than flour is used. The product tends to cook more evenly and to retain its form better once cooked. Because the yield of semolina is higher from *T. durum* varieties than from *T. aestivum* varieties of wheat, *durum* wheats have become the preferred raw material for all paste products. This preference has been converted into a definite requirement in the highly automated, industrial-scale production of pasta in the Western world. As well as its superior cooking qualities, *durum* semolina has another advantage to the industrial pasta producer in that it requires the addition of less water than any other wheat fraction in order to form a suitable paste. This is of great significance as pasta is normally dried for storage in the West rather than cooked at once.

5.6.6 Characteristics of *T. durum*

The *durum* wheats are relatively well adapted to semi-arid conditions and are grown principally in parts of North America, Argentina, the USSR, and in the Mediterranean countries. The yield tends to vary considerably but good *durum* wheat normally sells at a premium on the world market and so is usually an attractive crop in the drier areas of the world. There are many varieties, but as a group the kernels tend to be amber rather than red or white, relatively large, and extremely hard. The protein content is normally high but no more than in the hard red spring wheats when grown under similar conditions; arid conditions favour a high protein content. Despite the high protein content, flour from *durum* wheat is never as strong as that from *aestivum* varieties and normally has a much higher amylase activity and free sugar content. An extremely important characteristic of the *durum* wheats is that the endosperm has a relatively high carotenoid content and is distinctly yellow. This colour passes through into the macaroni products and is a distinguishing mark of pasta made from *durum* rather than *aestivum* wheats. Because of this, a yellow colour is associated by many people with pasta of a superior quality.

5.6.7 Criteria of quality of *T. durum*

As normal with any grain used for processing, *durum* wheat should be clean, free from seeds of other plants, including *T. aestivum* varieties, and the kernels should be of an even size. Apart from visual acceptability, good samples of *durum* wheat could be expected to have high test weights and high 1 000 corn weights. Although it is sufficient in some countries to assess hardness by external appearance, in others, such as the EEC, the sample corns have to be cut and the percentage of vitreous endosperms estimated. Within the EEC, *durum* wheat, as well as being at least 90 % derived from *T. durum* varieties, must contain at least 50 % vitreous kernels. There is also a requirement that the protein content should be above 13 %. Even where there is no legal limit, *durum* wheat is expected to have a high protein content and the milled flour should fall into the medium strong category.

The amount of yellow carotenoid pigments is normally determined by extraction of whole ground wheat with water-saturated *n*-butanol and should be as high as possible. The figure obtained in this way is not an entirely reliable guide to the eventual colour of the macaroni products for two main reasons. Firstly, the distribution of carotenoids between the endosperm and the rest of the kernel varies depending on the variety, growth conditions, and age of the wheat. Secondly, the wheat contains enzymes such as lipoxidase which destroy the carotenoids during processing. Lipoxidase activity in whole wheat can be estimated by an oxygen absorption test but, as about 80 % of the enzyme is in the embryo and bran, data from whole wheat only provide an approximate estimate of the activity likely to remain in the semolina. Obviously, lipoxidase levels should be as low as possible but careful processing is the key to maintaining macaroni colour. Short-term steam conditioning before milling can destroy lipoxidase completely, but the conditions are very critical if damage to the protein is to be avoided.

As with the *T. aestivum* wheats, milling behaviour and semolina yield are very important criteria of quality. Semolina is derived from the first break in the milling process (Fig. 5.4) and with *durum* wheats most of the extractable material can be obtained in semolina form. The particles in semolina normally range from 0.25 to 1.0 mm, whereas flour particles are smaller than 0.13 mm. The exact size of semolina required depends on the type of paste-forming process to be used. For hand-operated batch machinery, relatively coarse semolina is best (passing a 30-mesh US standard sieve but held on a 60-mesh sieve). Finer particles are preferred for automatic continuous processing (passing a 40-mesh sieve but held on an 80-mesh sieve). With increasing automation there has been a general move towards finer and finer semolina and some plants operate with an intermediate material, which is essentially a granular flour where all the particles pass through an 80-mesh sieve.

Whatever the exact size of semolina, it must be effectively grit-free and have a high protein and carotenoid content. As for the whole corn, lipoxidase activity should be low. Although instruments are available to measure the strength of the semolina paste and hand gluten-stretching tests can give useful information, test macaroni production is still required to assess a semolina reliably.

5.7 PEANUT PRODUCTS

Peanuts, or groundnuts as they are referred to in the feedstuff industry and in agriculture (Section 1.4), are one of the world's major crops and nearly 70% goes for oil production. They contain about 50% oil and the material produced is a good-quality cooking and salad oil (Section 5.14). The protein-rich residues can be used as a feedstuff. Peanuts are produced in the warmer parts of the world and on harvesting contain 18–25% moisture. As moisture contents over 13% allow mould growth, it is necessary to dry them rapidly to about 8% moisture to ensure safe storage. Low levels are necessary because any immature kernels may contain 4–5% more moisture than the mature kernels and so can easily become a centre of mould growth. Peanuts pick up off-flavours particularly easily from moulds, but even more important is that many of the fungi which grow on peanuts can produce toxins, e.g. strains of *Aspergillus flavus* produce aflatoxins and contaminated meal has been the cause of substantial poisoning amongst intensively reared livestock (Section 3.3). Although drying is important it is best effected in bins open to the air for several days, as temperatures much over 42 °C, which may occur in artificial driers, tend to detract from the flavour of the nuts once roasted. Properly dried peanuts can be kept for 5 years or more in cold stores.

Some peanuts are sold untreated 'in shell' for eating, but the bulk of the crop, which is not pressed for oil, is either roasted or converted into peanut butter. The exact process for the manufacture of peanut butter depends on the manufacturer but always includes the basic steps of shelling, dry roasting, blanching, and grinding. A number of machines, either batch or continuous, are available for dry roasting peanuts, but normally the shelled nuts are heated at 160 °C for 40–60 min. At first, moisture is lost from the peanuts and then as the oil cells are disrupted oil bursts through on to the surface—the so-called white roast. From this point on colour and flavour begin to develop (the dark roast). Once a satisfactory roast has been achieved the nuts are cooled rapidly in air and passed through a dry blanching process in which the skins, together with the radical plus shoot (heads), are removed by rubber brushes and belts. Peanut butter itself is produced by grinding the sorted, blanched nuts. This is normally a two-stage process with a medium grind first, followed by a second grind which gives a smooth paste. Other textures can be obtained by adding appropriately milled nuts to the paste before packaging. Considerable heat is generated during grinding and the machines are usually water cooled. The system is run such that during grinding temperatures of 60–85 °C are maintained, which ensure fluidity of the paste without causing burnt flavours to develop. Before filling into jars, other ingredients such as salt and fat stabilizers are added. The peanut butter is then cooled to between 30 and 43 °C to ensure proper setting in the jar. It is also important for this process that the peanut butter is deaerated before filling and that the filled jars are kept undisturbed for 2 days.

In the other major food use of peanuts, salted peanuts, the nuts can be dry roasted and then blanched, or blanched and then fried in oil. In either instance the process is adapted to produce the desired flavour and shine on the surface of the nut. About 2% of salt is added to the roasted nuts and adheres to the oily surface. It has been found that salt flakes are very much superior to granulated salt for salting peanuts. Another important point about the quality of the salt is that it must be low in metals, especially copper and iron, as these two catalyse oxidation reactions in the oil, which leads to rancidity.

The suitability of peanuts for both roasting and peanut butter manufacture is judged mainly by the absence of off-flavours and moulds and by the ability of the nuts to give the right type and degree of typical flavour on roasting. The basis of the roasted flavour seems to be compounds formed between amino acids and sugars, but the exact chemistry is not clear. In general, peanut roasters depend heavily on buying known varieties from known locations. Test roastings are then carried out before final quality can definitely be established.

5.8 COCOA AND CHOCOLATE

Cocoa and chocolate are both prepared from cocoa beans, the seeds of the tropical tree *Theobroma cacao* (Sections 1.5.6 and 1.6.4). Because of the sensitivity and short period of viability of cocoa beans, the first stage of processing, fermentation, has to take place locally in the plantation remote from the sites of the large cocoa manufacturing companies, which are mainly located in North America and Europe.

Within 1 day of being removed from the tree, the ripe pods are opened, normally with a knife, and the beans, together with much of the pulp, are scooped out. In the simplest fermentation process, the beans, still with large amounts of pulp attached, are made into a heap at once, covered with banana or plantain leaves and left for 5–6 days. Occasional mixing of the heap is necessary to maintain an even temperature throughout and, probably because of better temperature control, a more even fermentation is usually obtained when it is carried out in an open-sided box rather than in heaps. It is essential that only sound, uninfected beans are used. Fermentation is critical to the quality of the eventual cocoa and chocolate. Unfermented beans cannot be processed commercially and do not give rise to a chocolate flavour. Normally, the pulp becomes liquefied and drains away during the first day as the temperature in the heap rises. The temperature should not be allowed to rise much above 50 °C but needs to be near this level for several days. The course of fermentation is obvious to the eye in a change of colour from purple to brown and is accompanied by an increase in the moisture content of the cotyledons and a change from a dense to a more open structure. Although the process is normally referred to as fermentation and there is considerable yeast and bacterial activity throughout most of the time, dead beans do not give rise to a chocolate flavour. It seems that germination must begin in the early stages, although, of course, it does not go far because of the high temperature that is fairly quickly attained. Another factor which tends to kill the plant is acetic acid formed by fermentation of the sugars in the pulp. Much of this acid is absorbed by the cotyledons and results in an internal pH of about 5.0.

Once fermentation is complete, the beans are very susceptible to fungal infection, and must be rapidly dried. In many areas, the beans can be dried on trays in the open sun and this method is thought by many to give the best flavour. However, in some circumstances heavy rain prevents this simple process from being used and artificial drying has to be carried out. A number of systems are available for this but an important point is that the beans must not be exposed to any combustion products, as the fatty components easily absorb smokey taints. Moisture levels of 6.0–6.5% are ideal for storage, but at levels of 8.0% and above fungus is almost certain to develop very rapidly.

An additional hazard in the storage of cocoa arises from the fact that all of the beans have to make extensive journeys in ships, starting in the tropics and ending up in relatively cold latitudes. Apart from the obvious dangers of damage due to handling and the possibility of attack by insects and rodents, the change in temperature will result in condensation of water in the hold unless there is adequate air ventilation. Extensive damage to sacks of beans, particularly those near the side of the ship, can result.

Most cocoa beans are transported and stored in bags of about 140 lb as this suits the production side of the industry well. Bulk handling and storage are possible but care is needed to avoid cracking the beans. Large stores are normally fitted with internal baffle systems for this purpose.

5.8.1 Criteria of quality of cocoa beans

Most of the cocoa beans sold are in what is known as the 'bulk cocoa quality' range. Beans in this category should contain not less than 55% of the dry weight of the cotyledons (or nib) as cocoa butter (Section 5.8.2). The shells should not account for more than 12% of the weight of the beans and the average weight of each bean should not be less than 1 g. Not more than 12% of the beans should be outside the range of one third of their average weight above or below the mean. The consignment should be free of all foreign matter and damaged or shrivelled beans. It is essential that no mouldy beans are present and for this reason germinated beans where the germ has fallen out are very undesirable. The beans should be cut open to check that there is no internal mould and this test also allows an assessment of the degree of fermentation. The colour of the nibs should vary from purple–brown to full brown; nibs which are purple or slatey in colour indicate that fermentation was insufficient. Because of their high fat content cocoa nibs are particularly prone to pick up taints, e.g. smoke or fuel, and also to retain relatively high levels of insecticides. Beans which have a high moisture content should always be rejected.

The really important property of any batch of cocoa beans is the sort of flavour that will develop on roasting, and experimental roastings are necessary to check that flavour will be satisfactory.

A small amount of cocoa beans are sold as 'fine grade cocoa'. The criteria are as for bulk cocoa but values well above the minimum would be expected. However, the prime quality of fine grade cocoa is its flavour (Section 1.6.4). It is impossible to list fine grade cocoas in order of merit as merit is solely a subjective decision on the type of flavour desired, and the most important test is the taste test after roasting.

5.8.2 Processing of cocoa beans (Fig. 5.5)

Although bulk cocoa arriving from the producing countries is normally fairly free from foreign matter, the first stage of processing is still cleaning. This is normally carried out with an arrangement of screens, air lifts, and magnetic separators. The next stage is roasting. Various commercial roasters are available but most rely on hot air or heated drums with operating temperatures between 95 and 120 °C. The length of roasting can vary from 20 to 90 min and the exact conditions depend on the processers' experience and on the uses for the beans. Beans intended for cocoa powder are usually well roasted, whilst those for chocolate are normally only lightly roasted and the full flavour develops during further processing. The main purpose of

Figure 5.5   Flow diagram for cocoa and chocolate manufacture

roasting is to develop the typical chocolate flavour but the process also serves to make the shells brittle, thus facilitating their removal in the subsequent step. In this step, known as winnowing, the roasted beans are passed between two rotating knurled rollers set so as to crack, but not crush, the beans. The resulting pieces of nib and shell are separated by sieves and air lifts. Most of the nib can be isolated with only small contamination from shell and germ. However, because both shell and germ are difficult to grind they are undesirable in superfine cocoas and chocolates and so for this type of product their removal is a more important criterion of the winnowing process than is yield of nib.

Until the early part of the nineteenth century, cocoa was consumed by grinding the roasted nibs and then boiling with water. As the resultant drink is very fatty, some cereal flour or other absorbant material was normally added to increase the palatability of the drink. However, in 1828, Van Houten used a hydraulic press to remove excess fat and this is the basis of the modern process. The roasted nib is ground and, because of the heat generated by the mill, emerges as a fluid (cocoa mass or chocolate liquor). Once solidified this material can be sold as unsweetened chocolate for use in cooking. In the production of cocoa powder, however, the cocoa mass is subjected to hydraulic pressure (425 kg/cm^2) to squeeze out about half of the fat (cocoa butter). The hydraulic press has been replaced by solvent extraction in some companies. In this case the cocoa butter must be suitably refined before it can be used in chocolate manufacture. The remaining material is cast into cakes, then ground and sieved to form cocoa powder. This material is often described as natural cocoa as compared with cocoa made by alkalization. This process was also introduced by Van Houten and involves either treatment of the nib, or the cocoa mass, with alkali. The resultant cocoa powder has a darker, redder colour and a stronger taste than are otherwise achieved.

After a century of popularity cocoa is declining as a drink and its place is being taken, to at least some extent, by varieties of drinking chocolate. The basic composition of drinking chocolate is 70% sugar and 30% cocoa, together with a flavouring agent such as vanilla. Drinking chocolate is not just a mixture, however, as the ingredients have to be cooked together to obtain the 'instant' solubility in hot and even cold liquids which is one of the main attractions of drinking chocolate.

Eating chocolate, as we know it, was also an invention of the mid-nineteenth century and is essentially a concoction of cocoa mass, sugar, and cocoa butter. Various flavours can be added and in some cases concentrated milk is incorporated to give milk chocolate. In order to obtain the desired smoothness in eating chocolate the mixture has to be extensively mixed, rolled, and finally finished in a conche or melangeur. The quality of a chocolate depends very much on the exact process used, as flavour as well as smoothness develops during mixing.

Cocoa butter is a light yellow fat softening above 30 °C and melting sharply at 35 °C. It is composed of glycerides of stearic (34%), palmitic (24%) and oleic acids (38%), with small amounts of linoleic acid (2%) also being present. Cocoa butter, especially that obtained from alkalized cocoa mass, has too strong a flavour to be used in many chocolates, particularly milk chocolate. It is common to use deodorized cocoa butter, at least in part, but this material, although satisfactory from a flavour point of view, has inferior keeping properties in comparison with those of untreated expressed cocoa butter which contains natural antioxidants.

Fig. 5.5 shows that there is a direct connection between the production of chocolate and cocoa powder via cocoa butter. In many cases, the demand for chocolate is such that a substitute or replacement for cocoa butter is highly desirable. This has proved difficult to achieve as so many of the properties of chocolate depend on the properties of the fat. Cocoa butter is rare in having a substantial content of stearic acid. Hence natural substitutes have not been easy to find. A fat known as 'Coberine' has been prepared by fractionation of palm oil and has a glyceride composition almost identical with that of cocoa butter. This is a Unilever product and has been used successfully in some countries to replace deodorized cocoa butter. One natural fat, Illipe butter, extracted from seeds of *Shorea* species, can replace cocoa butter up to 10% but above this it affects both flavour and handling properties. Some synthetic fats can be tolerated up to 30% of the total cocoa butter. Many formulations exist in the chocolate industry but organoleptic and physico-chemical properties are always difficult to match to the natural product. In addition, legislation in many countries severely restricts the use of adulterants in chocolate.

5.8.3 Chocolate substitutes

It is known that cocoa butter is not important in the development of chocolate flavour during roasting and it seems likely that flavour is due to interactions between water-soluble amino acids, peptides, and sugars in the fermented bean. Artificial chocolate flavour can be produced by roasting milk protein hydrolysates and partially hydrolysed maize protein. Such products, when combined with deodorized cocoa butter and sugar, are used by the confectionery industry in the production of white chocolate and in a number of proprietary mixes.

5.9 SOYBEAN PRODUCTS

The soybean (*Glycine max*) is a plant which, after thousands of years of cultivation and domestic use in China and Japan, has suddenly become a major world crop in the third quarter of this century. It is an important export crop in the USA where it has been increasingly planted as cotton has declined. If dried to below 12% moisture the beans can be easily stored for 12 months without deterioration and, even if fungal attack does occur, production of mycotoxins is very unlikely. For marketing purposes the normal criteria apply to soybean, e.g. freedom from diseased and damaged seeds and absence of foreign material. Germinability is not a marketing criterion other than when sold for seed, but a good batch of soybeans would be expected to be 85–90% viable.

The beans themselves contain about 20% oil and 40% protein and, in the West, the bulk of the crop is processed to produce oil and a protein-rich oil cake for the feedstuffs industry. Good agronomic properties and ease of processing seem to have been the main driving forces behind the growth of the soybean industry as the oil has to be treated before it can be used in food (Section 5.14) and the protein, although of good quality, is deficient in methionine (Section 4.3). The bean has other drawbacks (Sections 4.3.4 and 5.15) which restrict its direct use in human diets, certainly as far as Europe and America are concerned.

In the Far East a variety of complex soybean products are important factors in the diet. The only one to have gained world-wide popularity is soy sauce or *shoyu*. This is a fermentation product of soybeans and wheat and the process is comparable to the production of sake (Section 5.5). Crushed, roasted wheat, cooked soybeans, and a *tane koji* containing *Aspergillus oryzae* and *A. soyae* are mixed to form a *koji* which, after addition of an 18% salt brine, forms *moromi*. Fermentation in the *moromi* proceeds through two stages: first a lactic acid fermentation and then a yeast fermentation. When fermentation ceases the liquid is filtered off, pasteurized, and bottled as *shoyu* (soy sauce). The pressed cake can be used as an animal feedstuff. The chemical composition of *shoyu* is very complex and includes a wide range of amino acids, organic bases, acids, and sugars. A rapid *shoyu* can be made by acid hydrolyis of defatted soybeans but it lacks the full taste of the fermented product and is commonly used as an extender.

A relatively similar process is used in the production of *miso*, except that much more variation occurs in the amount and type of cereal used, and after *moromi* the entire mash is blended to form a paste. The *miso* is not normally eaten by itself but is used as a base for a variety of soups.

Another soybean product popular in the Far East is *tofu*, which is a type of soy cheese. Soy milk is first prepared by cooking the soaked and ground beans and then filtering off the solids. A curd is then precipitated from the milk by addition of calcium sulphate. This salt curd is eaten directly without any microbiological ripening process as is commonly used with cheeses.

Along with *shoyu*, *miso*, and *tofu*, a wide range of fermented and non-fermented foods derived from soybeans are prepared both in the home and on a considerable industrial scale. For many people living in the Far East these products take the place of meat and dairy products. However, it seems that tastes and ideas would have to change considerably before people of European descent and culture became enthusiastic consumers of soybean products.

5.10 MUSTARD

In contrast to the cereals and some legumes, mustard is not a large-scale food crop but it is the most important native spice in the temperate regions of the world. Agriculturally it has several advantages as a break crop, especially between cereals, as it is easy to grow, yields well, and can usually be combined directly after the cereals have been gathered. As far as the UK is concerned, production runs at about 140 000 tonnes per year and most of the crop is grown in the traditional areas of East Anglia and the Fens. In world terms, the biggest areas of mustard production are in the South of Canada and the Northern USA.

Two types of mustard are commonly grown: brown mustard (*Brassica juncea*) and white mustard (*Sinapis alba*). These plants give rise to flours with different properties and commercial English mustards are produced by making suitable blends. Seeds are normally sown in the spring and the crop should be ready by early September in England. Once ripe, modern varieties of mustard do not lose their seeds, except under extreme weather conditions, so that the farmer has a little leeway in timing the harvest. The moisture content of the seed when harvested should not exceed 15%, especially when direct combining is used. For storage, even lower moisture contents are needed, and to be safe the level should be under 10%. In many cases this means the use of artificial driers. However, great care must be exercized to avoid excessive heating, which can result in serious loss of quality. Air temperatures over 65 °C and seed temperatures over 52 °C should be avoided. In this case it is not a matter of preserving viability, as with barley intended for malting, but rather it is important that the basic cellular structure and chemical composition of mustard be maintained right into the final flour.

Before any processing is carried out, the mustard seed must be free from all debris and weed seeds. Contamination can result in a gritty texture or visible flecking of the mustard flour at levels above 0.1%. As well as cleaning in conventional equipment, the purity of mustard seed can be conveniently improved by exploiting the fact that, because the seeds are small and spherical, they will rapidly move away from contaminants when allowed to roll down a slope. This process is known as 'spiralling' because of the shape of the equipment. However, although most contaminants can be removed in this way, there are two weed seeds, charlock in brown mustard and cleavers in white mustard, which cannot be separated. The only satisfactory policy is to eliminate these weeds from the land used.

Production of English mustard requires the prior production of mustard flour and the seed is milled in a process which is very similar to wheat milling (Section 5.6.3). The seeds are first dried and then moistened to the correct level before passing through the first break roller, which cracks the husk and allows the kernel to be separated cleanly. The kernel then passes through a series of reduction rollers to produce flour of the required grade. A major difference from wheat milling is that the cells of the mustard kernel must not be disrupted but merely separated. Otherwise, oil is liberated which results in a loss of yield as well as a loss of desirable chemical and physical properties of the flour. The fine level of control necessary is possible because mustard seeds are very uniform in size and so the distances between the rollers can be set very accurately. For this reason, brown and white mustard must be milled separately; 0.5 kg of brown mustard seed contains about 182 000 seeds, whereas 0.5 kg of the much larger white mustard contains only about 64 000 seeds. The separated husk, which comprises between 15 and 20% of the seed weight, is normally sent to an oil mill for solvent extraction of the oil which can be used both in industry and in food preparation. The residual oil-cake is used as an animal feedstuff (Section 4.3).

The organoleptic properties of mustard are due to the presence of the mustard oils, which are mainly colourless, volatile, irritating liquids with a pungent aroma. It has been known for a long time that the mustard oils are not present in the intact cell but exist as glycosides, and it is only on disruption of the cell that the enzyme myrosinase can act to liberate the mustard oil typical of the species. This is the main reason why processing temperatures must be kept low and cell disruption avoided. Brown and white mustards differ significantly in the mustard oils which they produce. Brown mustard gives rise to allyl isothiocyanate, a very pungent compound (equation 1), whereas white mustard gives rise to p-hydroxybenzyl isothiocyanate (equation 2). The latter compound is non-volatile, so white mustard has no pungency although it does have a characteristic hot taste. White mustard also tends to have higher levels of myrosinase than does brown mustard.

Equation 1

$$CH_2{=}CH.CH_2C \overset{\displaystyle NOSO_3}{\underset{\displaystyle S-glucose}{\Big\langle}} \xrightarrow[\displaystyle +H_2O]{\displaystyle myrosinase} CH_2{=}CH.CH_2.NCS$$

Sinigrin (in brown mustard)   Allyl-isothiocyanate

Equation 2

$$HO{-}\langle\bigcirc\rangle{-}CH_2C \overset{\displaystyle NOSO_3}{\underset{\displaystyle S-glucose}{\Big\langle}} \xrightarrow[\displaystyle +H_2O]{\displaystyle myrosinase} HO{-}\langle\bigcirc\rangle{-}CH_2NCS$$

Sinalbin (in white mustard)   p-Hydroxybenzyl isothiocyanate

The manufacturer of English mustard therefore blends brown and white mustard flours plus a certain amount of wheat flour and spices to give a product which, on mixing with water, will give rise to the desired level of pungency.

German mustard is normally made from white mustard alone and so is not pungent. In the traditional process a flour is not prepared, but the whole seeds including the husks are ground to a fine paste, the texture of which can be varied by the miller. This paste is then mixed with cereal flours, herbs, spices, water, and vinegar according to the recipe being followed. The mustard oil *p*-hydroxybenzyl isothiocyanate is formed during the milling, so the product is ready to use, unlike English mustard which has to be wetted to develop flavour. French mustards are made in a way similar to that used for German mustards except that normally only brown mustard is employed as a raw material. Again, because of the wet milling used, the mustard oil allyl isothiocyanate is formed at that point and, as much of it is lost because of its volatility, the final product is much less pungent than might be expected from the amount of brown mustard used. As with German mustards, a variety of French mustards are prepared by blending the mustard paste with a range of other ingredients.

Apart from its incorporation in mustard designed as a condiment, white mustard flour is used in several other ways in the food industry. It is a valuable ingredient in many salad creams as it has a good colour and emulsifying properties. It is also used as a meat extender. The separated husk, as well as being a source of oil, can also yield vegetable gums or can be used whole as cattle feed.

5.11 COCONUTS

The coconut (*Cocos nucifera*) is sometimes considered simply as one of the more important oil seeds, although its international significance has fallen with the introduction of large-scale hydrogenation of the unsaturated oils obtained from seeds such as soybean, peanut, and rape (Section 5.14). However, many other aspects of the coconut are important, e.g. the fibre, and it seems that almost no part of the fruit is not exploited at some level.

5.11.1 Copra

Copra is the dried endosperm or 'flesh' of the coconut (Section 1.6.2). Good copra can only be made from fully ripened nuts, which require a full year from the fertilization of the female flowers. The nuts can be picked by people on the ground using poles, but it is said that it is necessary to climb the tree and test the nuts individually before cutting, so as to ensure an absolutely even ripeness. The productivity of climbers is, of course, much lower than that of ground cutters. When 'ball' copra is being made the nuts are usually left to fall to the ground themselves to ensure maximum ripeness. Because the nuts are so large it is normal to quantify the crop by counting rather than by weight.

The first stage of processing is the removal of the exocarp and fibrous mesocarp. This is usually done by hand using a knife or a fixed blade of some kind. The hard 'coconut' remaining (as seen in food shops in the West) is then split in two by a single machete blow at its 'equator'. In the next part of the process the halved nuts are dried, either in the sun or in a kiln. For successful sun-drying long sunny spells and a low rainfall are necessary but, as these conditions do not favour the growth of coconut palms in the first place, sun-drying is relatively restricted in application. After 2–3 days in the sun the endosperm halves, or copra as they are now called, are detached from the shells (endocarp) and are dried for a further 4–5 days. Because of difficulties with the weather, a wide variety of kilns are in use to carry out the essential operation, which is to lower, fairly quickly, the moisture content of the coconut flesh from about 50% to less than 10% and preferably less than 6.5%. The prime criterion of copra quality is moisture content, as this is the major determinant of deterioration by bacteria, fungi, and insects. Other criteria are that the colour should be as white as possible, the copra halves should be large, thick, and smooth with a sweet smell, and, of course, the copra should be free of foreign matter or diseased pieces.

A small proportion of the copra production is eaten locally but most is ground and the oil extracted. Much of this process is carried out in the industrialized countries rather than the producer countries. Coconut oil is a highly saturated 'lauric' oil (Table 5.3) which finds many uses in the food industry and in soap and detergent manufacture. Locally in the producer countries the oil is also used for lighting.

The copra cake, or poonac, left after oil extraction is mostly used as an animal feedstuff.

With ball copra, mentioned above, the difference is that the nut is not split before the first stage of drying so that when the shell is eventually removed an intact, hollow ball of copra is obtained. This type of copra is used mainly for eating.

5.11.2 Desiccated coconut

In the production of desiccated coconut the shell and brown testa are chipped and pared away from the white endosperm, normally by hand although paring machines are available. The separated kernels are then sterilized in boiling water before being shredded or cut into fancy pieces by machine. This coconut meal is dried to about 2.5% moisture at a temperature near 70 °C and then packed in water-tight containers to produce the desiccated coconut used in food and confectionery and in home baking.

5.11.3 Coconut fibres (coir)

Coir is best made from slightly immature nuts, about 11 months old, which means that the complementary copra is not of the best quality. The main producers of coir are India and Sri Lanka. Indian production is typically of mat fibres which can be spun into ropes and yarns, whilst in Sri Lanka bristle and mattress fibres are produced. After collection of the nuts, the husk (exocarp plus mesocarp) is separated from the central nut as in copra production. In this case, however, it is the husk which is wanted. For the production of mat fibre in India the husks are soaked in ponds, backwaters, or specially dug pits for up to 10 months to allow retting of the tissue by hydrolytic enzymes (mainly pectinases) produced by a number of *Micrococcus* spp. which flourish in the water. It seems that the best results are obtained where the water is slightly saline and is changed periodically. Once retting is judged to be sufficient, the husks are removed from the water, washed, and the fibres beaten out with wodden mallets. The fibres are then dried, cleaned, and graded before being spun into a number of different types of yarn. The spinning is done largely by women working at a cottage-industry level.

In Sri Lanka, retting is carried out for a period between days and weeks. The husk pieces are then crushed between iron rollers and the fibres separated by use of revolving drums. The operator holds a piece of husk against the revolving face of a wooden drum equipped with protruding nails (the breaker drum). This tends to remove small fibres and pieces of connective tissue. The process is repeated with a second drum (the cleaner drum) which revolves more slowly and has thinner nails. The final result is that tufts of irregular bristle fibre are left in the operators hands. The minimum useful length for bristle fibre is considered to be 18 cm. The tufts of fibre are washed and dried before being bleached, dyed, and cut into standard lengths for sale. Some batches have the ends softened chemically and then shredded (flagging) to give a soft feel to the tip and improve sweeping efficiency.

The mass of material torn away by the breaking and cleaning drums is sifted through a series of wire-mesh cages to isolate the small mattress fibres, which are then washed and dried before curling and twisting. As the name implies, this grade of fibre is used for stuffing mattresses and similar jobs, but also finds application in some filtration systems and as an insulating and sound-proofing material.

5.11.4 Coconut shells

Coconut shells are a major by-product of copra manufacture and have a variety of uses. A good activated charcoal can be made which is suitable for gas absorption and bleaching in the chemical and food industries. The dried shells can be ground to a fine powder which is employed as a filler in some thermosetting plastics and is particularly useful where a good surface lustre is needed. One of the simplest, but very important, uses is as a convenient utensil in the producing areas, and to some extent the shells become a raw material for craftsmen producing decorated bowls, ornaments, and smoking pipes. Another simple local use is as a fuel, both domestically and in the kilns used for drying the nuts and the various products.

5.11.5 Coconut palm products

Although this chapter is about seed products it seems worthwhile to mention that the usefulness of the coconut palm is not limited to the production of nuts. It is common practice to tap the tree itself and collect the sugary sap (toddy), which can be drunk or used as the basic material for the production of palm sugar, either as a solid or as a treacle, fermented toddy, coconut vinegar, and arrack, a spirit distilled from fermented toddy.

5.12 COFFEE

Although there is no food value in the drink made from coffee beans, the flavour of the drink and possibly the stimulating effect of the alkaloid, caffeine, which is always present, has made it extremely popular over the whole world and today coffee is one of the most valuable international commodities.

5.12.1 Harvesting and preparation of coffee beans

Harvesting the ripe berries is a very prolonged and labour-intensive task as, like cocoa, the berries tend to ripen over a long period of time and, certainly for the best coffee, have to be picked by hand at intervals of a few days. Arabica coffee takes several weeks to ripen and once ripe the berries fall to the ground, so picking must occur about every 10 days so as to avoid substantial loss. Robusta berries take up to 11 months to ripen and do not fall spontaneously. Rapid harvesting techniques such as knocking ripe Arabica berries off with a stick or stripping off whole branches of Robusta berries tend to damage the berries and also produce a much wider degree of ripeness in the crop than does hand picking. Needless to say, the final coffee tends to suffer because of this.

The first stage of processing is the separation of the coffee beans from the surrounding skin and pulp of the berry. In very dry conditions such as occur in the Yemen, the berries can be dried in the sun. The dried pulp is crushed by passage through a stone mill and is then removed by winnowing. In wetter countries, or where production is large, a wet processing technique is usually employed. Wet techniques are also preferred for the best grade coffees but require expensive machinery and a copious supply of water. The dried, green beans are sorted to remove diseased beans and foreign material and batches of uniform size are collected. The dry, green beans are stored ready for the final part of the process which is roasting.

Roasting is necessary to develop the typical coffee flavour and must not be done too far in advance of use, as the aromatic oils (caffeol) begin to disappear as soon as they are formed. Whole roast beans can retain their flavour for several weeks but once ground the rate of loss increases and flavour can be greatly diminished within 7 days. A common roasting device is a directly heated rotating drum, but a number of techniques are used, varying in scale from hand-operated machines in a coffee shop to large industrial equipment. The extent of roasting, short of burning, is a matter of taste and distinct variations in preference occur in different countries. The time of day at which the coffee is consumed is also an important factor in determining the flavour required, e.g. mid-morning coffee may be lightly roasted whilst after-dinner coffee is often highly roasted and may be mixed with materials such as chicory to darken the colour and give a more bitter taste.

5.12.2 Instant coffee

Instant coffee is the main product made from coffee beans on a large industrial scale. The market for instant coffee has grown enormously since the 1940s and it is now a major way in which coffee is sold.

The process of instant coffee manufacture can be broken down into a number of stages, the first two of which are the direct equivalents of the stages in normal coffee production. The first obligatory process is roasting. As well as the conventional roasting techniques, instant coffee manufacturers may also use a variety of methods of roasting, including continuous processes, high-pressure roasting and roasting in a fluidized bed. An important point in all roasting is to control the quenching of the hot beans so that roasting is quickly brought to a stop at the appropriate time but without producing significant amounts of steam. This reduces the loss of aroma volatiles. The extraction processes used are essentially similar to the domestic processes of coffee making. In most cases the freshly ground coffee beans are mixed with water at the optimum volume and temperature for extraction and flavour development. The process is repeated with the freshly extracted grounds and then repeated once or twice more but using higher temperatures and pressures. Many variations of the exact conditions used exist and some techniques are more akin to the use of a percolator or expresso machine than to the jug method.

The part of the process unique to instant coffee is, of course, the removal of water from the freshly made coffee extract to yield the coffee solubles in a dry but freely dissolvable form. Two main types of technique are in commercial use: spray drying and freeze drying. In spray drying the extract is heated to near its boiling point and then sprayed out as a fine mist into the top of a drying chamber through which cooling air is passed. The dried coffee collects in the bottom of the vessel as a fine powder. Because of the high temperature used, spray drying inevitably results in flavour changes in the coffee, although this effect is utilized to advantage in the preparation of high roast instant coffee as it means that the preliminary roasting of the green beans can be shortened. The alternative process of freeze drying is currently gaining in popularity because, as it occurs at low temperatures, changes in the taste of the coffee are avoided. For freeze drying a completely clear coffee extract is required, so a filtration step is normally necessary after extraction. The next part of the process is usually a freeze concentration step in which much of the water is frozen out as a sludge of fine ice crystals. The concentrated coffee extract is then frozen completely and the water removed by flash volatilization at low temperature in a vacuum. The conditions can be manipulated so that the resultant instant coffee occurs as small particles similar in appearance to ground coffee rather than as a powder. This type of appearance seems to be a selling point and a number of agglomeration techniques have been invented to convert instant coffee powder into granules.

However it is made and whatever its appearance, the main drawback of instant coffee is the almost complete loss of the volatile fraction, and so the coffee aroma, during manufacture. A variety of approaches designed to restore some aroma are in use. All seem to involve an initial removal of the essential oil from roasted beans before extraction with hot water and then addition of the aroma concentrate as a frost or solution to the instant coffee before packaging. Apart from the additional cost of such 'aroma enrichment' the process seems to have an adverse effect on the keeping quality of the instant coffee, presumably owing to rancidity developing in the lipid fraction.

5.12.3 Decaffeinated coffee

Decaffeinated coffee is prepared in the same way as conventional instant coffee except that the caffeine is extracted from either the green beans or the roasted beans before further processing. The commonest solvents are chlorinated hydrocarbons such as trichloroethylene and dichloromethane, although other solvents such as wet supercritical CO_2 and glycerol triacetate can be used. As with instant coffee, the main problem in manufacture of decaffeinated coffee is loss of flavour components during processing.

5.13 STARCH

The main source of starch throughout the world is maize, although in some countries other plants, e.g. potato, sago, tapioca, and wheat, may also be important.

The first stage in the extraction of starch from maize is to soak the grain in water until a moisture level of 50% has been achieved. The liquid (corn steep liquor) is drained away and the moistened grain carried in a stream of water through an attrition mill, usually with one stationary and one rotating steel plate, to yield a slurry of broken endosperm pieces, pericarp, and embyro. The germ has a relatively low specific gravity, because of its high oil content, and can be separated from the rest of the slurry in flotation chambers or in the smaller and more efficient hydrocyclones which depend on a whirlpool effect. The separated germ is normally dried and sent for oil extraction. The remaining slurry is then passed through a second mill, which reduces the particle size of the starch and gluten but has little effect on the pieces of fibre, which are subsequently removed by filtration. The starch particles are heavier than the gluten particles and can be separated using a continuous-flow centrifuge. After drying to 10% moisture the gluten material contains about 70% protein, whilst after two or three washes the protein content of the starch slurry (39% starch) can be reduced to 0.3%.

Starch cannot be produced so easily from wheat because of the unique properties of wheat gluten, and several different techniques are used for the preparation of wheat starch. Under modern conditions wheat starch is best viewed as a by-product of wheat gluten manufacture as it is this product which makes the processing worthwhile.

Although not a seed, it seems worth mentioning that starch of a high purity (0.05% protein) can be obtained from potatoes. The process is commonly used in The Netherlands, where special potato varieties are grown for the purpose. One disadvantage is that potato starch tends to retain a distinct flavour. Likewise, sago starch is not derived from seeds but rather from the pith of a palm which is grown in the Far East and, although important locally, accounts for only a small proportion of world starch production. The procedure for separation of maize starch can also be successfully applied to manufacture of starch from damaged and small grains of rice and to a number of types of sorghum.

Starch is a product with an extremely wide variety of uses and it is common practice now to adjust the properties of the native starch by chemical modifications so as to obtain a superior product. For example, while untreated starch can be used as a paste many of the difficulties of solubilization and dispersion can be eliminated if the starch has been gelatinized first by a suitable heat treatment. This results in so-called 'instant' or cold-water pastes. Treatment with 2-diethylaminoethyl chloride results in a much stronger paste which can therefore be used more sparingly. A similar situation exists with regard to the use of starch in the food industry where the fluidity, setting properties, and clarity of the starch gels can all be adjusted to suit the product by a range of substitution and cross-linking techniques.

Apart from the use of starch as starch, there is a major industry based on the hydrolysis products of starch. Syrups with a wide range of sugar contents can be produced from starch by acid or enzymic hydrolysis for use as sweeteners, as cereal substitutes in the fermentation industries, or as the raw material for caramel manufacture. With the drive beginning in the late 1970s at first to dilute and eventually to replace petrol with alcohol, in such countries as Brazil and the Philippines, it seems possible that one day the alcohol-for-drinks industry may be dwarfed by the alcohol-for-fuel industry. The source of sugar for this type of fermentation is determined by availability and cost rather than any other criterion and at the moment, although no seeds are being used on an appreciable scale, extensive use is being made of sugar cane and cassava (tapioca), a tropical plant with large starchy roots. The sugary extract from sugar cane can be used directly for fermentation but the cassava starch has to be converted to fermentable sugars by the use of added microbial enzymes. Although the need for conversion of the starch increases the cost of using starchy materials for fermentation compared with sugar cane, the better harvesting and storage properties of the starchy plants are a compensating factor.

The list of uses of starch and starch-derived products is extremely long. As well as the examples mentioned above, these materials find extensive use in the pharmaceutical, textile, and paper industries. They are also used in the production of specialized products such as foundry moulds and artificial muds used during the drilling of oil wells, particularly under the sea.

5.14 OIL

About 40 plant species are used as sources of vegetable oil at the moment although there are hundreds of other species which could be used. The exact pattern of world oil production is continuously changing but a limited number of seeds account for about 90% of trade. These are soybean, sunflower, peanuts, cottonseed, rapeseed, coconut, palm and palm kernel, linseed, and castor beans. The last two give rise to non-edible, entirely technical oils whose main use is in paints, lubricants, and as raw materials in plastic production. The other oils are mainly used as salad oils, in cooking, and for hydrogenation to produce margarine. Although oil quality is obviously an important criterion, the commercial availability of an oil often depends more on the non-oil uses of the seed rather than on the need for the oil itself. A good example of this is the soybean. The seed contains only about 18% of oil (Table 5.3) but, because of the demand for soybean protein, soy oil is by far the commonest vegetable oil available today. In a similar way, but on a much smaller scale, corn oil is available really as a by-product of the maize starch industry and the supply of cottonseed oil is a reflection of the demand for cotton.

Olive oil and sesame oil represent two special cases as, although trade in these two is fairly small, their joint production is substantial and is nearly equivalent to the production of rapeseed oil. Effectively, olive oil and sesame seed oil are largely consumed in the Mediterranean countries where they are produced.

Although each type of seed necessitates different detailed extraction and purification procedures, the same principles and general approach apply to all of them. In general, the oil is obtained either by direct physical pressure from a hydraulic or screw press often referred to as an expeller, or by solvent extraction, normally with hexane. In some cases it is beneficial to combine both approaches. The solid residue, the oil cake or oil meal, can be dried and used for a variety of purposes (Section 4.3).

Before extraction it is essential that the seeds or, in cases such as maize, the separated embryos, are clean and free from broken or contaminated seeds or weed seeds. Foreign matter which remains in the oil cake can substantially reduce the value of this product, and unwanted or other lipids can arise from contaminated seeds or weed seeds. It is also common practice to dehull, or decorticate, seeds before extraction, as this tends to improve both oil and oil cake quality. However, in some cases where the seeds are very small, e.g. rape, the husk cannot be removed without excessive loss of oil.

Immediately before extraction the seeds are roughly crushed in roller mills and then steam cooked at temperatures in the range 85–95 °C. Cooking fulfils a wide variety of functions, from breaking down the oil cells and precipitating phosphatides to inactivation of undesirable enzymes, e.g. myrosinase in rape. It is also critical that, after cooking, the seeds have the right moisture content for the ensuing extraction process. Physical pressing is most economical when applied to seeds with a high oil content. The disadvantage of this system is that the oil cake is likely to contain considerable amounts of oil, perhaps up to 10 %, which represent a large loss in extraction over the theoretical maximum and endow it with poor keeping qualities. Furthermore, excessive pressing tends to reduce the nutritive value of the oil cake because the heat generated denatures the protein and selectively destroys amino acids, particularly lysine. With solvent extraction, although it involves an extra step in solvent removal, the oil can be almost totally extracted. A common technique used with seeds of high oil content is to press relatively gently to leave an oil cake with about 20 % oil and then recover this by solvent extraction.

The crude oils obtained are likely to contain a wide range of compounds apart from triglycerides. These include free fatty acids, di- and monoglycerides, phosphatides, sterols, gums, waxes, vitamins, pigments, proteins, and a wide variety of sulphur compounds and oxidation products. These can normally be removed by a suitable combination of filtration, precipitation, acid and alkali treatment, bleaching, and steam-stripping to give an acceptable refined oil. In many cases, chemical antioxidants are added to the refined product to prevent deterioration due to formation of free fatty acids. Mono-*tert*-butylhydroquinone (TBHQ) is commonly used for this purpose at concentrations up to 200 ppm.

The quality of an oil can only be defined in terms of its intended use. Unlike the situation with unfractionated products, this relationship can often be closely defined in chemical terms, as can the keeping qualities. The following paragraphs consider closely some of the special points which apply to the major oil seeds.

Data on oil composition tend to be variable and there seems to be no doubt that a number of factors are known to influence oil composition. As well as analytical error, these include the growth environment of the seeds, their condition on harvesting, the extent of drying, the conditions of storage, and the method of oil extraction and storage. Despite this variability the composition of the major oils tends to show some reproducibility and it is possible to make some general observations (Table 5.3). The fatty acids which predominate are palmitic, oleic, and linoleic. Stearic and linolenic acids are minor components and are often absent completely. Some oils contain large proportions of relatively rare fatty acids, e.g. ricinoleic acid in castor bean and erucic acid in rapeseed oil. Most oils contain a number of minor components which may or may not contribute to oil quality. In the hard oils derived from palm and coconut, the whole range of saturated fatty acids from C_8 to C_{18} tend to be present in substantial amounts.

Table 5.3. Fatty acid composition and total oil content of the major vegetable oils (mainly from Hilditch and Williams, 1964). (Reproduced by permission of Associated Book Publishers Ltd.) Most oils contain small amounts of other fatty acids. The major saturated fatty acid is palmitic, normally with minor amounts of stearic. However, in the seeds with high levels of saturated fatty acids the whole range from C_8 to C_{18} is present and lauric acid (C_{12}) may account for up to 50% of the total, with the second most important saturated acid being myristic acid (C_{14})

Botanical name	Common name	Fatty acids (%)				Total oil content (% dry wt.)
		Saturated	Oleic	Linoleic	Linolenic	
Linum usitatissimum	Linseed	6–16	13–36	10–25	30–60	30–40
Carthamus tinctorius	Safflower	5–10	13–37	57–59	0	20–38
Helianthus annus	Sunflower	9–17	14–72	33–72	0	20–40
Gossypium hirsutum	Cottonseed	12–25	20–44	33–54	0	18–25
Arachis hypogaea	Peanut	9–30	40–66	20–38	0	38–50
Zea mays	Maize (embryo)	8–16	23–49	34–56	0	24–32
Glycine max	Soybean	10–18	22–30	50–60	5–9	13–24
Brassica compestris,						
B. napus	Rape seed*	4–14	14–29	12–24	1–10	35–40
Ricinus communis	Castorbean†	1–2	1	4	0	40–55
Olea europea	Olive	10	83	7	0	12–15
Elaeis guineensis	Palm kernel	75–85	10–18	1–2	0	50
Elaeis guineensis	Palm (mesocarp)	40–45	39–47	5–11	0	30–40
Cocos nucifera	Coconut (copra)	90	5–8	1–2	0	63

*Contains 40–54% erucic acid [docos-13-enoic acid, $CH_3(CH_2)_7CH=CH(CH_2)_{11}COOH$].

†Contains 92–94% ricinoleic acid [12-hydroxyoctadeca-9-enoic acid,

$$CH_3(CH_2)_5CHOHCH_2CH=CH(CH_2)_7COOH].$$

Although the properties of an oil depend on the exact mixture of triglycerides present, the fatty acid composition is normally very revealing. As unsaturation increases so the melting point decreases. This tends to be accompanied by increased fluidity at room temperature and the ability to remain clear at low temperatures, points which are of importance in salad oils, especially when refrigerators are in common use. The reverse generalization applies to the smoke point and flash point of an oil, although in these cases the exact composition and stability of the triglycerides are particularly important. Since the di-unsaturated fatty acid linoleic acid may be essential in the diet, the most desirable food oils are generally considered to be those with a high oleic and linoleic acid content. Although any polyunsaturated fatty acid can be oxidized by the air, the susceptibility increases with degree of unsaturation and so linolenic acid is much more prone to autoxidation than is linoleic acid. Linolenic acid is normally considered an undesirable component in a food oil because of this tendency, which can result in changes in the chemical and physical properties of the oil and in the development of off-flavours. The significant linolenic acid content of soybean oil lowers the value of this product in comparison with most of the other food oils. Although chemical hydrogenation can be used to reduce the linolenic acid, the process is expensive and also tends to remove the desirable linoleic acid at the same time. The specificity of the catalytic process has recently been improved by the use of copper–chromium catalysts, which avoid substantial chemical reduction of linoleic acid. Oil partially hydrogenated in this way has to be treated by bleaching and winterization (a chilling and filtering process) to reduce the copper levels, but there seems to be no difficulty in obtaining acceptable concentrations in the final oil. A more satisfactory answer to the problem would be to develop new strains of soybeans whose oil lacks linolenic acid in the first place but whose other properties are at least up to those of the varieties in use at the moment.

Whilst autoxidation has to be avoided in the food oils, the process is essential for so-called drying oils. These oils, which include linseed oil, are added to paints and on exposure to the atmosphere oxidize to form a tough, dry film over the surface. A good drying oil should contain at least 70% polyunsaturated fatty acids, with preferably over 50% linolenic acid. From the results in Table 5.3, linseed oil would appear to be the only suitable drying oil. However, safflower, sunflower, and soybean oils can come into the category of 'semi-drying', as in many batches from some localities the concentration of linoleic acid is sufficiently high to bestow drying properties. Olive oil is an example of a completely non-drying oil.

As well as being used directly in food and its preparation, a considerable amount of vegetable oil is hydrogenated to produce fats which are the basis of margarine. Here the criteria for use, as well as general availability and price, tend to be the absence of unwanted flavours and the ability to produce a fat with physical properties equivalent to those of butter. Considerable quantities of vegetable oils are used in the production of soups and detergents and as rubber extenders. There are many minor industrial uses which depend on the general properties of oil, but in a few cases there is a demand for one of the rarer fatty acids as a basic material. For example, both erucic acid and ricinoleic acid can be used for the production of nylon. Ozonolysis of erucic acid (equation 3) yields pelargonic acid and brassylic acid. Pelargonic acid can also be made from oleic acid and finds application in a wide range of products from plasticizers to synthetic flavours and jet-engine lubricants. Brassylic acid is also used as a plasticizer but a potentially more significant

Equation 3

$$CH_3(CH_2)_7 CH = CH (CH_2)_{11} \cdot COOH + O_3 \longrightarrow CH_3(CH_2)_7 \cdot COOH$$

Erucic acid   Pelargonic acid

$$+ HOOC \cdot (CH_2)_{11} COOH$$

Brassylic acid

function is that it can be used to produce amino derivatives which are the monomers for several types of nylon.

Because of its long carbon chain, erucic acid can also be used to form waxes which are suitable bases for such products as floor polishes and shoe creams.

5.15 PROTEIN

Compared with the production of oil, and even starch, the amount of protein extracted from seeds is relatively insignificant. Some use is made directly as a food additive, but even in the case of a massive crop such as soybeans, only a few per cent of the seed is handled in this way. In most cases a flour or protein-enriched fraction is simply mixed in with a cereal flour at a low rate, to improve the nutritional quality, but without upsetting the flavour or handling qualities of the cereal product. One distinctive change in recent times, however, has been the attempt to produce textured vegetable proteins in direct competition to meat and meat products. Soybean protein has been the main raw material. Much industrial secrecy still surrounds the manufacturing processes but basically there appear to be two main types. The more important is a spinning technique in which the prepared, heated protein mixture is forced through an orifice to form a fibre. The fibrous mass can then be shaped in a secondary process to give the appearance, and hopefully texture, of chunks of meat or mince. Most of the research effort seems to have been directed at achieving a mouthfeel which is indistinguishable from that of meat by empirical means, and little is known at a basic level. Apart from the emotional aspect, major shortcomings of textured vegetable proteins are the persistence of the rather unpleasant soybean flavour and the inadequacies of artificial meat flavours.

The second type of process produces sheets of textured protein and is used for the simulation of bacon and other sliced meats. The prepared protein mixture is cast on to a roller to form a thin sheet and this is removed from the roller by a blade in such a way as to cause constant crinkling and crimping. This results in a sheet of material with a certain amount of meaty texture. In some situations prior texturizing of the soybean protein does not seem to be necessary. For example, when it is used as a sausage filler, the protein sets to form a gel which has mouthfeel properties similar to those of a finely minced or puréed meat filling.

The preparation of soy proteins themselves on the commercial scale is a relatively simple process, and again two main industrial processes are used. In the first, defatted flakes are extracted with water and the soluble proteins precipitated by acidification to about pH 4.5. There are many variants of this so-called 'acid-leach' process. The alternative approach is to extract carbohydrates and low molecular weight compounds from the defatted flakes with alcohol and alcohol–water mixtures and so leave behind a protein-rich material. In most manufacturing processes these soybean protein preparations are not used alone but are mixed with various flours, oils, eggs, and milk products as well as the appropriate flavourings and stabilizers.

Soybean protein has a high nutritional value and, so long as soybean products are well cooked to destroy potentially dangerous components such as trypsin inhibitors, agglutinins, and saponins, are a valuable addition to the range of human food (Section 4.3.4). Flavour seems to be a major obstacle to obtaining further popularity and here, of course, price becomes a very important factor. Production of animal protein is a very inefficient process and therefore the product is costly compared with the feedstuff. The production of textured proteins seems worthwhile only if the cost of manufacture is distinctly less than the cost of animal production.

Apart from direct food uses, seed proteins have a number of minor industrial uses. For example, the gluten derived from cereals has a very high glutamic acid content and can be hydrolysed to yield a mixture of amino acids which is an effective enhancer of meaty flavours. A more specialized use of a gluten preparation is the coating of a variety of nuts used in confectionery with zein. This is the gluten derived from maize and, by forming a transparent film over the nut, improves both the appearance and the keeping qualities. Protein preparations from groundnuts can be spun into wool-like fibres and also form the basis of glues used in the manufacture of plywood.

Uses of this type seem likely to expand in the future as mineral oil, the basis of so many synthetic materials, becomes rarer and increases in price.

REFERENCES

General references

Appelquist, L. A. and Ohlson, R. (1972). *Rapeseed*, Elsevier, Amsterdam.

Ashby, H. K. (1977). *Cocoa, Tea and Coffee*, Priory Press Ltd., Hove, Sussex.

Brouk, B. (1975). *Plants Consumed by Man*, Academic Press, New York.

Child, R. (1974). *Coconuts*, 2nd ed., Tropical Agriculture Series, Longmans, London.

Jarman, G. G. and Jayasundra, D. S. (1975). *The Extraction and Processing of Coconut Fibre*, (G94), Tropical Products Institute, London.

Knight, J. W. (1969). *The Starch Industry*, Pergamon Press, Oxford.

Minifie, B. W. (1970). *Chocolate, Cocoa and Confectionery*, Longman, London.

Pintauro, N. D. (1975). *Coffee Solubilization—Commercial Processes and Techniques*, Noyes Data Corp., Park Ridge, N.J.

Pomeranz, Y. (1971). *Wheat, Chemistry and Technology*, 2nd ed., American Association of Cereal Chemists, St. Paul, MN.

Pomeranz, Y. (1976). *Advances in Cereal Science and Technology*, American Association of Cereal Chemists, St. Paul, MN.

Roden, C. (1976). *Coffee*, Faber, London.

Rose, A. H. (1977). *Alcoholic Beverages*, Academic Press, New York.

Rosengarten, Jr., F. (1969). *The Book of Spices*, Livingston Publishing Co., Wynnewood, PA.

Smith, A. K. and Circle, S. J. (1972). *Soybeans: Chemistry and Technology, Vol. 1, Proteins*, Avi Publishing Co., Westport.

Tooley, P. (1971). *Fats, Oils and Waxes*, John Murray (Publishers) Ltd., London.
Urquhart, D. H. (1961). *Cocoa*, Longmans, London.
Wolf, W. J. and Cowan J. C. (1975). *Soybeans as a Food Source*, CRC Press, Cleveland, Ohio.
Woodroof, J. G. (1973). *Peanuts: Production, Processing, Products*, 2nd ed., Avi Publishing Co., Westport.
Woodroof, J. G. (1974). *Coconuts: Production, Processing, Products*, 2nd ed., Avi Publishing Co., Westport.

Specific references

Hilditch, T. P. and Williams, P. N. (1964). *The Chemical Constitution of Natural Fats*, 4th ed., Chapman and Hall, London.
Spicer, A. (1975). *Bread*, Applied Science Publishers, Barking.

INDEX

150